1988

University of St. Francis
510.711 A330
Albers, Donald J.,
Undergraduate programs in the

3 0301 00074927 1

SO-AYH-420

UNDERGRADUATE PROGRAMS
IN THE MATHEMATICAL
AND COMPUTER SCIENCES
The 1985-1986 Survey

Donald J. Albers
Richard D. Anderson
Don O. Loftsgaarden

LIBRARY
College of St. Francis
JOLIET, ILLINOIS

Supported by the National Science Foundation under grant S R S 8511733

Any opinions, findings, conclusions, or recommendations expressed herein do not necessarily reflect the view of the National Science Foundation.

This survey has been carried out under the auspices of the **Conference Board of the Mathematical Sciences** (CBMS). This Conference Board of organizations provides a means to address matters of common concern to the mathematical community in its broadest sense. The Conference Board is located at:

1529 Eighteenth Street, NW
Washington, DC 20036

CBMS Member Organizations

American Mathematical Association of
    Two-Year Colleges
American Mathematical Society
American Statistical Association
Association for Symbolic Logic
Association for Women in Mathematics
Association of State Supervisors of
    Mathematics
Institute of Mathematical Statistics
Mathematical Association of America
National Council of State Supervisors of
    Mathematics
National Council of Teachers of Mathematics
Operations Research Society of America
Society of Actuaries
Society for Industrial and Applied Mathematics
The Institute of Management Science

CBMS Chairman: F. Joe Crosswhite

CBMS Survey Committee

Donald J. Albers, Menlo College, Chair
Richard D. Anderson, Louisiana State
    University, Executive Director
Kim B. Bruce, Williams College
William G. Bulgren, University of Kansas
Wendell H. Fleming, Brown University
Barbara Gale, Prince George's Community
    College
Don O. Loftsgaarden, University of Montana,
    Data Analyst
Donald Rung, Pennsylvania State University
Joseph Waksberg, WESTAT Research
    Corporation

Copyright 1987 by
The Mathematical Association of America
Library of Congress Catalog Card Number 87-061084
ISBN 0-88385-057-5

Current print (last digit)
  6 5 4 3 2 1

510.711
A 330

PREFACE

At five-year intervals, beginning in 1965, the Conference Board of the Mathematical Sciences (CBMS) has, with the financial support of The Ford Foundation in 1965 and later the National Science Foundation, conducted surveys of undergraduate programs in the mathematical and computer sciences as found in universities, four-year colleges, and two-year colleges. The surveys have obtained much information on undergraduate course enrollments, faculty, and teaching patterns in mathematical and computer science departments. The basic purpose of these surveys has been to provide information useful for decision-making in mathematical and computer science departments, professional organizations, and government agencies. The surveys have reflected the interests of the members of CBMS* and have drawn heavily on the expertise and experience of prominent individuals from the various areas of the mathematical and computer sciences represented by these organizations.

All five CBMS surveys have addressed two basic questions:

1. What are the national undergraduate course enrollments in mathematics, statistics, and computer science, how are those enrollments distributed among various types of institutions of higher education, and how do the enrollment patterns change over time?

2. What are the numbers, qualifications, personal characteristics, and teaching responsibilities of faculty in the mathematical and computer sciences, and how do these variables change over time?

* CBMS members are listed on the facing page and on the back cover.

Publisher #9.35

3-10-88

128,217

In addition to establishing trend data on these basic questions, the present survey has initiated four special new thrusts deemed to be of importance in the mid-80's:

1. Much more detailed identification of faculty, course, and student phenomena in computer science.

2. Identification of faculty who are teaching computer science while they are members of mathematical science departments. This is especially important when examining questions related to organization of mathematical science departments and deployment of mathematical science faculty.

3. More detailed information on remediation.

4. Identification of various issues judged to be important by departments.

Questionnaire design and overall advice and guidance for the present Survey of Undergraduate Mathematical and Computer Science were provided by the CBMS Survey Committee. The members of that Committee are as follows:

Donald J. Albers, Menlo College, Chairman
Richard D. Anderson, Louisiana State University, Executive Director
Kim B. Bruce, Williams College
William G. Bulgren, University of Kansas
Wendell H. Fleming, Brown University
Barbara Gale, Prince George's Community College
Don O. Loftsgaarden, University of Montana, Data Analyst
Donald Rung, Pennsylvania State University
Joseph Waksberg, WESTAT Research Corporation

We very much appreciate the help of Robert Aiken, Chair of the Education Board of the Association for Computing Machinery, in identifying computer scientists Kim Bruce and William Bulgren to serve on the Survey Committee and in reviewing a draft of the report.

The work of survey sample design, data analysis and report writing has been shared by three people. Data analysis and design of the sampling and estimation procedures was chiefly the work of Professor Don O. Loftsgaarden, who also was a member of the 1980 CBMS survey project. In the early stages of sample design, Professor Loftsgaarden was assisted by Joseph Waksberg, an internationally known figure in this area of statistics.

The writing of the present report has been primarily the work of the undersigned. For several years in the 1970's Professor Anderson directed survey programs of the American Mathematical Society. Professor Albers, Chairman of the present Survey, largely authored the chapters on mathematical and computer sciences in two-year colleges in the 1975 and 1980 reports, as well as in the present one. In addition to designing the questionnaires for the present survey, the members of the Survey Committee reviewed the draft of the report making many helpful comments.

CBMS and its Survey Committee are indebted to Maureen Callanan of the Mathematical Association of America and to the other MAA staff members who administered and supported this project. Special thanks and appreciation for grant support are due the National Science Foundation, which also supported CBMS's 1970, 1975, and 1980 surveys.

Our special thanks to Cherie C. Wilks for preparation of the final manuscript.

Donald J. Albers
Chairman, Survey Committee

Richard D. Anderson
Executive Director of the Survey

TABLE OF CONTENTS

TABLE OF CONTENTS (Continued)

TABLE OF CONTENTS (Continued)

TABLE OF CONTENTS (Continued)

TABLE OF CONTENTS (Continued)

TABLE OF APPENDICES

LIST OF TABLES & GRAPHS

LIST OF TABLES & GRAPHS

CHAPTER TWO

LIST OF TABLES & GRAPHS

CHAPTER TWO (Continued)

LIST OF TABLES & GRAPHS

PAGE NO.

LIST OF TABLES & GRAPHS

LIST OF TABLES & GRAPHS

LIST OF TABLES & GRAPHS

CHAPTER SIX (Continued)

INTRODUCTION

We present results from the 1985 CBMS Survey. In general, the data show for the period 1980-85 that undergraduate mathematics has recovered some ground lost in the seventies and that undergraduate computer science grew very rapidly in the first half of the eighties. There is an increasing quantification of many facets of society demanding more young people with knowledge of mathematical topics and with the ability to use computers to address increasingly complex problems of society. Thus people trained in mathematical thinking seem to be in increasing demand. The longer term enrollment data support this contention. The rather rapid development, over the past fifteen years, of computer science as an undergraduate academic discipline shows considerable adaptability of the educational system to powerful external forces. But the rather limited and slow changes observed within undergraduate mathematics itself show less impact of the forces of change.

We list Survey results in the university and four-year college sectors (Chapters 1-4) separate from those in the two-year college sector (Chapters 5-6). We include a special section on undergraduate programs in the computer sciences as Chapter 4, the result of a special questionnaire on computer science.

The Survey Committee feels, as its counterparts in the past have felt, that it should present data and findings without much policy interpretation. The factual background given here is for the use of those in education and science policy positions to use in making informed decisions. Thus we deliberately avoid making recommendations on policy issues, leaving such activities to people or groups responsible for making policy.

Our findings concern mathematical and computer science enrollment trends, faculty characteristics, instructional formats and administrative organization. The data given are estimates of national totals for fall 1985 in institutions of higher education. The estimates are based on responses to questionnaires sent to a stratified random sample of schools from among 2,463 institutions with undergraduate programs in the mathematical or computer sciences. The stratification was by total student enrollments in universities, four-year colleges and two-year colleges. The sampling and estimation procedures are explained in Appendix A. The table given later in this introduction shows sampling and response rates in various categories of institutions and departments. The consistently high response rates in various strata give us confidence in the overall data reported although the lower response rates from computer science departments make the details of computer science data somewhat less reliable. See Appendix F for the list of all respondents. The lists and categories of universities, public four-year colleges, private four-year colleges and two-year colleges were obtained from NCES (National Center for Educational Statistics, now the Center for Education Statistics) lists of the most recent year (1982) available to us at the time of preparing the sample. Similar lists were used in the 1980 Survey. It should be noted that the list of universities is not the same as that used in the annual AMS Survey of doctoral producing departments. There is an overall 70-75% overlap with the AMS lists and a larger percentage overlap with AMS Groups I and II departments. Generally the four-year public college category is comparable to, but larger than, the AMS masters producing department category (M) and the private four-year college category is comparable to the AMS bachelors producing category (B) but they are not, in fact, identical and considerable variation from AMS data is to be expected. The Survey Committee felt that the advantages of using lists comparable to those of the 1980 Survey and of those used in other disciplines outweighed the advantages of using AMS lists. Indeed, we sampled institutions, not departments, in order to get national characteristics. For the institutions in the sample, questionnaires were sent to all mathematics departments or to the division in charge of mathematics courses. In addition, questionnaires were sent to all computer science, statistics or other mathematical science departments that were determined to exist at sampled institutions.

This Survey provides a valuable statistical data base concerning what was going on in the fall of 1985 in collegiate mathematics, statistics, and computer science and what changes have occurred over the previous five to twenty-five years. It should be of continuing value to educational policy makers in and out of the mathematical and computer science communities. But the reader should keep in mind that it is not designed to give more than background information on important issues facing our community and our increasingly technological society. Among these issues not specifically addressed by the Survey are:

- What should our youth be learning to equip them (and us) to face the challenges of an ever-more rapidly changing technological world of tomorrow and how well equipped is our system and our faculty to address these developing student needs?

- What are the support levels and mechanisms necessary to effect a transition into education for the 21st century?

If anything, the data appear to suggest both rather slow adaptation to a rapidly changing society except in the development of computer science as an undergraduate discipline. Support levels in the mathematical sciences seriously lag even existing patterns of change.

SAMPLING AND RESPONSE RATES IN DEPARTMENTS
OF MATHEMATICS, STATISTICS AND COMPUTER SCIENCE

	Population	Sample	Respondents	Response Rates
Universities				
Mathematics	157	72	56	78%
Statistics	40	21	19	90%
Comp. Sci.	105	51	32	63%
Public 4-Yr. Colleges				
Mathematics	427	105	81	77%
Statistics	5	2	2	100%
Comp. Sci.	141	40	24	60%
Private 4-Yr. Colleges				
Mathematics	839	80	57	71%
Comp. Sci.	150	16	8	50%
Two-Year Colleges	1040	172	110	64%

The response rates generally were a bit higher than those for the 1975 Survey and a bit lower than those for the 1980 Survey. We believe the responses (compared for early and late respondents) are generally adequate to justify the numerical conclusions given.

There are two major periodic surveys in the mathematical community - (1) the CBMS Survey conducted every five years (with Ford Foundation support in 1965 and with NSF support since 1970) published in a form such as this report, and (2) the annual American Mathematical Society (AMS) Survey, with reports published periodically in the AMS Notices. Both Surveys are directed by committees appointed by the sponsoring professional organizations. Over the years, the two committees have had considerable overlapping membership. The committees actively cooperate with each other and compare data.

The CBMS Survey is much lengthier and more detailed and is based on careful statistical sampling (with followups) and with projections to the total populations. It is concerned primarily with undergraduate education, is designed to cover both the mathematical and the computer sciences, and in both 1980 and 1985 has been based on lists of undergraduate institutions available from the Department of Education. It has a related but separate component on two-year colleges.

The AMS Survey is primarily a faculty and new doctorate survey, it concentrates now only on the mathematical sciences (since the response rates from computer science departments were getting progressively worse), and it gets limited information on course enrollments as a byproduct. However, since the AMS gets comparable data from both the current and past year, it monitors year-by-year changes very effectively.

The categories of institutions used by the CBMS Survey are Universities (Public and Private for sampling, but since 1980 reported in one "University" category), Four-Year Public Colleges, Four-Year Private Colleges and Two-Year Colleges. The AMS Survey currently classifies departments in the mathematical sciences by Groups I, II, and III (PhD producing mathematics departments), Group IV (Statistics departments), Group V (Applied mathematics, OR, etc., departments with doctoral programs), Group VI (Canadian departments) and Group M (Masters producing) and Group B (Bachelors producing) departments.

Both surveys collect and present data by their categories of institutions on numbers of full- and part-time faculty, on numbers of GTA's, and on enrollments in selected types of courses. Since the categories do not explicitly correspond, the numbers by categories can not be directly compared. However restricting the CBMS data to university mathematics departments, and after allowing for known differences in the specific institutions on the AMS and CBMS lists, the figures for numbers of full-time faculty and GTA's are in close agreement, with the AMS totals being expectedly about 10% higher than the CBMS totals.

Another relevant survey is the so-called Taulbee Survey of the Computer Science Board. Like the AMS Survey it is concerned primarily with PhD programs and their graduates and with the faculty of such programs. The 1985-86 Taulbee Survey had responses from 117 out of 118 PhD producing computer science departments (including 10 Canadian departments). The total faculty size in these 107 U.S. departments was almost half-again as large as that shown for the full-time faculty in the departments identified by CBMS in the university category (which included a number of computer science bachelors or masters producing departments). It is believed that this difference is largely explained by the known variations in the lists of departments in the categories used in the two surveys, by the fact that the Taulbee Survey figures apparently included some or all part-time faculty (presumably on a pro-rata basis), and different possible interpretations of whether to count visitors and/or faculty on leave.

SUMMARY HIGHLIGHTS

We give some of the highlights of the Survey as a summary of the results. The reader is advised to note carefully the distinctions made at the beginning of Chapter 2 with respect to various components of the faculty. The reader is advised to read the relevant portions of Chapters 1 to 6 to better understand the limitations or qualifications of these highlights.

HIGHLIGHTS

FOUR-YEAR COLLEGES AND UNIVERSITIES

- While overall undergraduate enrollments in universities and four-year colleges were almost stable since 1980, mathematics course enrollments increased by 6% to 1,619,000, statistics by 41% to 208,000, and computer science by 74% to 558,000.

- The number of undergraduate degrees in mathematics and statistics (all types including mathematics education) was 20,096, up from 13,906 in 1980 but not up to the 24,181 level of 1975. The number of degrees in computer science was 29,107, up from 8,917 in 1980 and from 3,636 in 1975.

- After a sharp rise from 1975 to 1980, the enrollments in remedial mathematics were 251,000, up from 242,000 five years earlier, a 4% increase.

- The enrollments in upper division mathematics courses were up 52% over 1980 levels reversing a downward trend from the '70's.

6

- Undergraduate statistics enrollments have been increasing markedly since 1960.

- The full-time faculty in the mathematical and computer sciences grew by 21% in the period 1980-1985 and now numbers 22,194 while the part-time faculty grew by 46% to 9,189.

- Since 1970, the FTE (full-time equivalent) faculty of all those teaching in the mathematical sciences in four-year colleges and universities increased by 6% while the course enrollments in the mathematical sciences increased by 41%. In the same period, the overall FTE faculty in the mathematical and computer sciences grew by 40% while course enrollments grew by 72%.

- The total computer science faculty (i.e. teachers of computer science) is now 5,651 full-time (3,605 in computer science departments), 5,342 part-time (1,984 in computer science departments) for a total 7,432 FTE, up from an estimated 1,182 FTE in 1970. See page 37 for explanations of special faculty terminology.

- The percentage of doctorates among the full-time faculty has decreased from 82% in 1975 to 73% in 1985 in the mathematical and computer sciences.

- Teaching load assignments generally are similar to those in 1970. Typical computer science and statistics faculty teaching assignments are less than those for mathematics faculty.

- The creation of new computer science departments and the broadening of departmental duties and names to include computer science were frequent administrative changes.

- In five major introductory courses, 41% of university students are taught in large lecture-type sections (over 80 students) whereas in private colleges only 2% are. About one-fifth of all students in these five courses are taught in sections of 40 to 80 students.

7

- There is little required use of computers in college algebra or calculus or in any mathematics course other than numerical analysis or other computing related courses.

- Since 1980 the number of graduate teaching assistants has been stable in university mathematics departments but has gone up markedly in statistics and computer science departments and in public college mathematics departments.

- About 95% of all graduate teaching assistants in mathematics, statistics or computer science are students in the same or related subjects.

- Salary levels and departmental support practices are widely regarded as major problems in mathematics and statistics departments.

- Two-thirds of all universities, one-third of all public four-year colleges, and more than one-sixth of private four-year colleges have separate computer science departments. In the private four-year college category the number is 150, more than triple that for 1980.

- Of the 3,754 doctorates who teach computer science full-time, 1,291 have their degrees in computer science and 1,555 in mathematics. Of the 2,231 doctorates who teach computer science part-time, 181 have their degrees in computer science whereas 1,369 have their degrees in mathematics.

- Half of all part-time computer science faculty teach full-time in the same institution, almost a third are employed outside education and a tenth are not employed full-time anywhere.

- Half (49%) of all computer science sections are taught in mathematics departments, the rest in computer science departments.

■ In a substantial number of institutions, some computer science is taught outside mathematics and computer science departments, chiefly in business, engineering and education academic units.

■ Total reported enrollments in computer science have climbed from 107,000 in 1975 to 321,000 in 1980 to 558,000 in 1985.

■ There were 29,107 computer science undergraduate degrees in fiscal year 1984-85, with 8,646 of these in mathematics departments. In addition there were 3,084 joint majors with mathematics. The number of computer science degrees reported in the 1980 Survey for fiscal year 1979-80 was 8,917.

■ About two-thirds of all institutions with computer science major programs require calculus for computer science majors, one-half require linear or matrix algebra and more than two-fifths require discrete mathematics.

■ The most common problems reported by computer science departments are salary levels and patterns, departmental support services, the need to use temporary faculty, and the upgrading and maintenance of computer facilities.

TWO YEAR COLLEGES

■ Mathematical science enrollments remained essentially unchanged since 1980, decreasing by 1% whereas overall two-year college enrollments decreased by 2%. Part-time students continued to account for nearly 2/3 of all two-year college students. Nearly 2/3 of all two-year college associate degrees are in occupational programs.

9

- Some specific mathematics course areas showing enrollment increases since 1980 in two-year colleges were statistics (29%), calculus (13%), remedial (9%), other precalculus (4%) and computing (3%). Some showing decreases were technical mathematics (56%), business mathematics (42%) and mathematics for liberal arts (42%). Remedial mathematics now accounts for almost 47% of all enrollments in the mathematical and computer sciences, up from 42% in 1980.

- The figures on course enrollments above are from mathematics programs per se. A substantial number of mathematics and computing courses are taught outside these mathematics programs. Estimates indicate that in 1985 more than 50% of business and technical mathematics was taught outside, about 20% of arithmetic, about 15% of statistics, about 80% of data processing and about 60% of computer science other than data processing. These courses were taught primarily in business and occupational programs.

- Access to computers as well as the impact of computers and calculators on mathematics teaching has increased. But even so, excluding computer science sections, less than 7% of all sections involve computer assignments for students.

- Mathematics labs have been established in 82% of all two-year colleges, up from 68% in 1980.

- Since 1980 there has been a marked decrease in the number of two-year colleges using any of the various alternative forms of instruction: TV, film, programmed, CAI, PSI, etc.

- Two-year college mathematical science faculty increased by 12% since 1980 in each of the full- and part-time categories. In 1985 there were 6,277 full-time and 7,433 part-time faculty. The percentage of doctorates among full-time faculty decreased to 13%, the first decrease noted since 1970. Since 1975 the percentage of women on the full-time faculty increased from 21% to 31% and the percentage of ethnic minorities increased from 8% to 12%. About 43% of the full-time faculty reported teaching overloads but overall teaching loads decreased for the first time since 1970.

■ Remediation was cited as the biggest problem facing faculties in the mid-1980's.

CHAPTER 1

UNDERGRADUATE STUDENTS

This chapter reports estimated national student enrollments in university and four-year college mathematical and computer science courses in fall 1985. Detailed course-by-course enrollments for universities, public four-year colleges and private four-year colleges are given in Appendix E. This chapter also contains analyses of undergraduate degrees granted in the mathematical and computer sciences. Extra computer science data is provided in Chapter 4. The current chapter provides some specially prepared data on undergraduate statistics. Where data is known and relevant, it also provides information on changes in undergraduate student phenomena over time.

HIGHLIGHTS

- While overall undergraduate enrollments in universities and four-year colleges were almost stable since 1980, mathematics enrollments increased by 6%, statistics by 41% and computer science by 74%.

- The number of undergraduate degrees in mathematics and statistics (all types, including mathematics education) was 20,096, up from 13,906 in 1980 but not up to the 24,181 level of 1975. The number of degrees in computer science was 29,107, up from 8,917 in 1980 and from 3,636 in 1975.

- After a sharp rise from 1975 to 1980, the enrollments in remedial mathematics were 251,000, up from 242,000 five years earlier, a 4% increase.

- The enrollments in upper division mathematics courses were up 52% over 1980 levels reversing a downward trend from the '70's.

- Undergraduate statistics enrollments have been increasing markedly since 1960.

TRENDS IN UNDERGRADUATE EDUCATION

We begin with some data over time from Department of Education Publications "Projections of Educational Statistics", "Digest of Educational Statistics" and other national compilations of information about undergraduates or prospective undergraduates. With this data as background, we then look at mathematical and computer science student data from this Survey. For the reader's convenience, we have organized much special data about computer science as a separate and later Chapter 4.

Since 1970, full-time undergraduate enrollments in all higher education (including two-year colleges) have increased by 20% and part-time enrollments have more than doubled. Overall FTE (full-time equivalent) enrollments have increased by 30%. Much of this increase has been at the two-year college level. Graph 1-A gives the full-time, part-time, and FTE enrollments over time.

GRAPH 1 - A

UNDERGRADUATE ENROLLMENTS IN HIGHER EDUCATION SINCE 1970
(In Thousands)

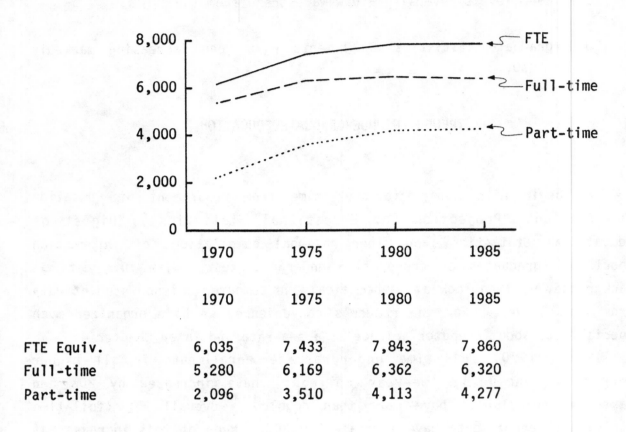

	1970	1975	1980	1985
FTE Equiv.	6,035	7,433	7,843	7,860
Full-time	5,280	6,169	6,362	6,320
Part-time	2,096	3,510	4,113	4,277

Based on reports from various institutions, Department of Education sources use a part-time student enrollment as equivalent to 36% of a full-time enrollment. Over the period 1970-1985, the total increase in overall FTE undergraduate enrollments was 30%. By comparison undergraduate student course enrollments in the mathematical and computer sciences in all of higher education increased by 76% from 1970 to 1985. Even with all computer science enrollments deleted, the increase in undergraduate enrollments in mathematical sciences from 1970 to 1985 was over 50%. And 1970, the base year, was at the end of a boom period in science in the 1960's. These figures clearly show a rapidly increasing

14

role for both the mathematical and computer sciences in higher education.

Looking only at the four-year college and university sector, overall FTE enrollments increased about 16% in the period from 1970 to 1985 and mathematical sciences enrollments (not counting computer science enrollments) increased 40%, (From Tables 1-2 and 1-10). This occurred over a period when almost the entire growth of the combined mathematics and computer science faculty since 1970 has been concentrated in computer science (Table 2-5).

PROBABLE MAJORS OF ENTERING FRESHMEN IN HIGHER EDUCATION

Table 1-1 below shows the trend over time of the choices of academic majors in a number of disciplines. The data comes from The American Freshman: National Norms for Fall 1985 by Astin, A. W., King, M. R. and Richardson, G.T. and earlier editions of this report. The trends in the various disciplines shown seem to conform to conventional wisdom. It is encouraging that the "mathematics and statistics" category appears to have "bottomed out". Among the "hard" sciences and engineering, only the mathematical sciences show an upswing, albeit mild, since 1980.

TABLE 1 - 1

PERCENTAGES OF ENTERING FRESHMEN PLANNING
MAJORS IN SELECTED DISCIPLINES

	1965	1970	1975	1980	1985
Business	14.3	16.2	18.9	23.9	26.8
Education	10.6	11.6	9.9	7.7	7.1
Engineering	9.8	8.6	7.9	11.8	10.7
Humanities & Arts	24.3	21.1	12.8	8.9	8.3
Mathematics & Statistics	4.5	3.2	1.1	0.6	0.8
Physical Science	3.3	2.3	2.7	2.0	1.6
Social Sciences	8.2	8.9	6.2	6.7	7.6
Computer Science	-	-	-	2.5	2.3
Data Proc. and Comp. Prog.	-	-	-	2.4	2.1

The 1986 figures for computer science and for data processing and computer programming were 1.9% and 1.6%. The profiles on "first choices of intended specific fields of study of college bound seniors" prepared annually for the College Board and involving responses from some million high school seniors show somewhat similar patterns and trends. In mathematics and statistics, the figures from 1975, 1980 and 1985 are 2.4, 1.1 and 1.1 respectively. However, the overall computer science and systems analysis figures for the same years were 2.8, 4.2 and 7.2.

TOTAL MATHEMATICAL AND COMPUTER SCIENCE ENROLLMENTS SINCE 1970

Graph 1-B gives total undergraduate enrollments in the mathematical and computer sciences in two-year colleges and in the four-year colleges and universities. The growth can be compared to that of all undergraduate enrollments shown in Graph 1-A and to that of faculty growth shown in Table 2-2.

GRAPH 1 - B

TOTAL MATHEMATICAL AND COMPUTER SCIENCE ENROLLMENTS IN HIGHER EDUCATION
(In Thousands)

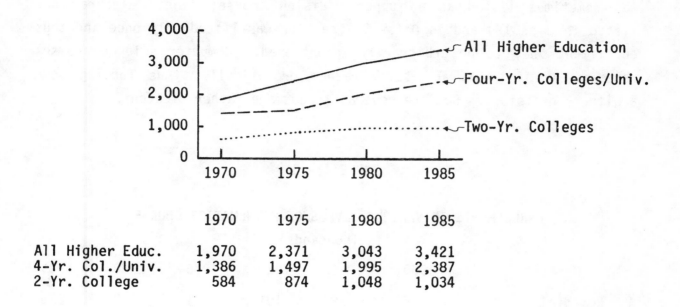

	1970	1975	1980	1985
All Higher Educ.	1,970	2,371	3,043	3,421
4-Yr. Col./Univ.	1,386	1,497	1,995	2,387
2-Yr. College	584	874	1,048	1,034

UNDERGRADUATE MATHEMATICS ENROLLMENTS OVER TIME

From the earlier Surveys and a Department of Education Report of 1960 authored by Clarence Lindquist (who also did the basic statistical work for the 1965 to 1980 Surveys) we see some interesting changes over time in undergraduate enrollments. We look first at mathematics course enrollments by levels of courses and separately at statistics and computer science. In Table 1-2 we give course enrollments by four categories A: Remedial (courses 1-4); B: Other pre-calculus (courses 5-14); C: Calculus level (courses 15-19); and D: Advanced (courses 20-44). The comparable long term trend data for statistics is in Table 1-11 and for computer science is in Table 4-10. See Appendix B or E for the course numbers and titles. We use the course designations and numbers from the

17

present Survey and adapt the course lists from the earlier Surveys to fit the present list. We used the lists on page 28 of the 1975 Survey report for the 1960-1970 data. It is necessary to make some arbitrary decisions, e.g. general mathematics (basic skills, operations) is regarded as a remedial course, A, even though it was not previously listed that way; mathematics for elementary school teachers is regarded as B even though it was sometimes listed as an upper division course; linear algebra (now listed in C as 19) and in D (as 34) earlier was listed only once and thus enrollments had to be arbitrarily apportioned. Computer science courses have changed in name and level rather dramatically, thus forcing some arbitrary decisions. But the general trends are rather clearcut.

TABLE 1 - 2

ENROLLMENTS IN VARIOUS LEVELS OF MATHEMATICS COURSES
(in Thousands)

		1960	1965	1970	1975	1980	1985
A:	Remedial	96	89	101	141	242	251
B:	Other pre-calc.	349	468	538	555	602	593
C:	Calculus level	180	315	414	450	590	637
D:	Advanced	92	133	162	106	91	138
	Total	717	1005	1215	1252	1525	1619

Roughly speaking, A represents high school mathematics taught in college, B represents other freshmen level mathematics at a level below calculus, C represents the first two years of mathematics for those able to start with calculus, and D represents upper division mathematics. It should be noted that a great deal of elementary statistics and computer science is also taught in mathematics departments. Thus figures in Table 1-2 and in Table 1-3 below do not represent departmental teaching loads but levels of mathematics courses taken. Below we give the percentages of mathematics courses taken at various levels over time obtained from Table 1-2 above.

18

TABLE 1 - 3

PERCENTAGES OF ENROLLMENTS IN VARIOUS LEVELS OF MATHEMATICS

		1960	1965	1970	1975	1980	1985
A:	Remedial	13%	9%	9%	11%	16%	15%
B:	Other pre-calc.	49%	46%	45%	44%	39%	37%
C:	Calculus level	24%	29%	34%	36%	39%	39%
D:	Advanced	14%	16%	12%	9%	6%	9%

There has been a small but encouraging increase in the sum of C and D from 45% in 1965 to 48% in 1985. The big jump in remedial enrollments for 1975-80 occurred at a time of the development of specially funded federal programs designed to get colleges and universities to address remediation issues and was accompanied by an equally large reduction in the percentage of enrollments in other pre-calculus mathematics. In that light, this change was merely a shift downward from other pre-calculus courses to remedial--perhaps a reflection of both falling student entrance test scores at the lower levels and more faculty attention to that problem.

Since most undergraduate statistics courses taught in mathematical science departments are taught in the (primary) mathematics department, it is reasonable to look at the total mathematics and statistics undergraduate load over time. The detailed (and explosive) growth in statistics enrollments, per se, is given in Table 1-11. Combining mathematics and statistics enrollments in the two categories of (1) pre-calculus and (2) calculus-and-beyond we have the following phenomena.

TABLE 1 - 4

COMBINED MATHEMATICS AND STATISTICS ENROLLMENTS BY LEVEL

	1965	1970	1975	1980	1985
Pre-Calculus Courses 1-14, 45, 46	54%	54%	57%	57%	54%
Calculus and Beyond 15-44, 47-54	46%	46%	43%	43%	46%

The 1960 figures were 61% and 39% making the detailed data from that original study somewhat suspect in light of this almost constant distribution of course load by level.

Since computer science as a subject has developed only within the past 25 years, there has, of course, been phenomenal growth in computer science enrollments over that period. The time trends for computer science are given in Table 4-10.

In Table 1-5, we give the enrollments in four-year colleges and universities over time in several specific mathematics courses.

TABLE 1 - 5

ENROLLMENTS OVER TIME IN SOME SPECIFIC MATHEMATICS COURSES
(in Thousands)

Subject	1960	1965	1970	1975	1980	1985
Arith./Gen. Math.	48	29	23	32	63	45
H.S. Alg. & Geo.	48	60	78	109	179	202
Lib. Arts Math.	36	87	74	103	63	59
Math for Elem. Teachers	23	61	89	68	44	54
Coll. Alg., Trig.	235	262	301	259	345	352
Finite Math.	1	7	47	74	95	88
Anal. Geo. & Calc.	184	295	345	397	517	534
Diff. Equations	29	31	31	29	45	45
Linear/Matrix Alg.	4	19	47	28	37	47
Adv. Calc.	17	20	20	14	11	14
Other Undergrad. Math.	(94)	(134)	(160)	(139)	(126)	(179)
Total Math. Enrollment (Stat. & C.S. not included)	717	1005	1215	1252	1525	1619

UNDERGRADUATE ENROLLMENTS IN THE MATHEMATICAL
AND COMPUTER SCIENCES FOR 1980 AND 1985

In Table 1-6A, we give 1980 and 1985 enrollments for various course levels in mathematics, statistics, and computer science and in Table 1-6B we give the separate totals for all undergraduate mathematics, statistics and computer science in these years.

TABLE 1 - 6A

1980 AND 1985 MATHEMATICS, STATISTICS AND COMPUTER SCIENCE ENROLLMENTS
BY LEVELS IN UNIVERSITIES & PUBLIC & PRIVATE FOUR-YEAR COLLEGES*
(In Thousands)**

	1980				1985				
	Univ.	Pu.	Pr.	Total	Univ.	Pu.	Pr.	Total	Ch.
Remedial math.	63	151	28	242	56	155	40	251	+4%
Other pre-calc.	214	261	127	602	200	280	113	593	-1%
Calc. level	282	175	133	590	281	258	101	637	+8%
Adv. level math.	28	29	32	91	47	66	25	138	+52%
Elem. stat./prob.	33	48	23	104	52	54	39	144	+38%
Adv. stat.	25	13	6	43	37	18	10	66	+53%
Lower level C.S.	69	77	60	206	94	155	101	350	+70%
Middle level C.S.	12	14	8	35	18	34	13	66	+89%
Upper level C.S.	30	35	19	80	54	61	28	142	+78%
Total	756	803	434	1993	839	1081	470	2387	20%

The enrollment figures above show that remediation is still a major but not a significantly growing problem. The increase in advanced level math enrollments was fairly evenly spread over all types of courses: core math, math for secondary school teachers and applied math.

The statistics figures are for enrollment in the mathematical and computer sciences type departments not in psychology, education, business, etc.

The list of computer science courses did not include data processing per se (at an elementary level) but a small number of data processing enrollments might have appeared in an "other" category.

* and ** See footnotes on next page.

TABLE 1 - 6B

TOTAL 1980 AND 1985 UNDERGRADUATE ENROLLMENTS IN MATHEMATICS, STATISTICS AND COMPUTER SCIENCE*
(In Thousands)**

	1980				1985			
	Univ.	Pu.	Pr.	Total	Univ.	Pu.	Pr.	Total
Mathematics	587	616	320	1525	584	759	279	1619
Statistics	58	61	29	148	89	72	49	208
Computer Science	111	126	85	322	166	250	142	558
Total	756	803	434	(1993)	839	1081	470	(2387)

*   It should be noted, as remarked in the Introduction to this report, that enrollments as well as faculty data in the university, public college and private college categories are not directly comparable to the AMS Survey Groups I, II, & III; M; and B categories. The Dept. of Education lists of institutions for the three categories from which the Survey samples were drawn have considerable but not total overlap with the AMS lists of departments. A comparison of the Survey and AMS lists suggest that total mathematics enrollments in the Survey "university" category should be marginally lower than enrollments in Groups I, II, & III departments.

**   The course-by-course enrollments are given in Appendix E. To maximize the accuracy of primary published data, they were individually rounded to the nearest thousand. This process led to some total enrollments being different from the sum of the addends, e.g. 1.3 + 2.3 + 3.3 = 6.9 rounds to 1 + 2 + 3 which is not 7. Consequently, the numbers in Tables 1-6A and 1-6B do not always sum correctly to the last digit.

AVAILABILITY OF SELECTED UPPER LEVEL MATHEMATICAL COURSES
IN UNIVERSITIES AND FOUR-YEAR COLLEGES IN 1985

In the 1985 questionnaire, departments were asked to report on whether particular courses were being offered in the academic year 1985-86 or <u>had been offered in the academic year 1984-85</u>. In previous surveys, the question did not contain the reference to the preceding year. The Survey Committee felt that because many advanced courses are only offered on a two-year cycle, particularly in smaller institutions, the proper reference frame on availability should cover a two-year cycle. It turned out that with this revised wording asking for the availability of courses over two years, there were much higher percentages of institutions offering various upper level courses. The Survey committee believes that this year's data more accurately represents the status of course availability. Twenty-one out of the thirty percentages below are about half again as high as those reported in 1980.

TABLE 1 - 7

PERCENTAGE OF INSTITUTIONS OFFERING SELECTED COURSES
IN 1984-85 OR 1985-86

	Course	Univ.	Pu.4-Yr.	Pr.4-Yr.
1)	Theory of Numbers	65%	56%	20%
2)	Combinatorics	63%	22%	5%
3)	Foundations of Mathematics	27%	30%	17%
4)	Set Theory	33%	24%	3%
5)	History of Mathematics	42%	39%	9%
6)	Geometry	79%	77%	47%
7)	Math. for Sec. Sch. Teachers	45%	55%	40%
8)	Mathematical Logic	35%	19%	12%
9)	Applied Math./Math. Model.	51%	37%	26%
10)	Operations Research	44%	33%	26%

AVERAGE SECTION SIZE AT VARIOUS COURSE LEVELS

From the main questionnaire on course enrollments and numbers of sections, we are able to get the following information:

- The average section size in remedial mathematics is about 32 with intermediate algebra sections a bit larger and arithmetic and general mathematics sections a bit smaller.

- The average section size in other pre-calculus mathematics is 35 with each course having an average section size within 3 of that number except for business mathematics with 43, finite mathematics with 39, and mathematics for elementary school teachers with 29.

- The average section size in calculus-level courses is 34 with calculus for biological, social and management sciences at 40 and discrete mathematics and linear/matrix algebra just under 30.

- The average section size for advanced level courses in mathematics is 19.

- In statistics, at the elementary (freshman) level the average section size is 37 and at the advanced level is 30.

- For the lower, middle and upper level courses in computer science the average section sizes are 31, 26, and 22 respectively.

College of St. Francis Library
Joliet, Illinois

BACHELORS DEGREES IN THE MATHEMATICAL AND COMPUTER SCIENCES

Three different types of data are given in the tables below: in Table 1-8, the overall numbers of bachelors degrees in various specialties for the twelve months ending on June 30 of 1975, 1980 and 1985; in Table 1-9, the 1984-1985 numbers of bachelors degrees by type of department; and in Table 1-10, the 1984-1985 bachelors degrees reported by <u>mathematics</u> departments and tabulated by category of institution. Together these tables and accompanying comments give an interesting picture of undergraduate major programs. The reader is also referred to Table 1-13 and Table 4-11 for separate data relevant to statistics and computer science degree programs.

The numbers given below include only the given institution's majors in mathematics, computer science or statistics departments (by whatever name it is called). There were eight other mathematical science departments of various special descriptions whose data were submitted in the Survey. But the total number (eight) of such departments divided among various strata for sampling was too small to make meaningful projections to national totals of undergraduate degrees for such types of departments. Those eight departments reported a total of 320 degrees. Thus the counts of degrees given in this Survey may be a bit low, particularly in some of the specialty areas.

The numbers of bachelors degrees in the mathematical and computer sciences took a major leap in the five year period from 1980 to 1985, with computer science degrees more than tripling and, when joint majors are included, overall mathematics degrees increasing toward the 1974-75 levels. The current Survey asked for counts of joint majors as well as for individual majors for the period July 1984 to June 1985. In earlier Surveys, such joint majors would presumably have been counted as degrees only in the field of the department in which they studied.

The recent 1985-86 Taulbee Survey of the Computer Science Board indicates a cessation of growth in the number of computer science undergraduate degrees. Recent AMS Surveys indicate a modest reduction in computer science enrollments.

College of _____ Emory Library

TABLE 1 - 8

NUMBERS OF BACHELORS DEGREES

Special Area	1974-75	1979-80	1984-85
Mathematics (General)	17,713	10,160	12,102
Applied Mathematics	1,120*	1,527*	1,215
Math. Education	4,778	1,752	2,567
Computer Science	3,636	8,917	29,107
Statistics	570	467	538
Operations Res.	---	---	312*
Joint C.S. & Mathematics	---	---	3,084
Joint Math. & Statistics	---	---	121
Joint C.S. & Statistics	---	---	157
Total	27,817	22,823	49,203

* The applied mathematics categories in 1974-75 and 1979-80 include figures from the small categories "actuarial science" and "other" not included in this year's questionnaire. However, the additional "operations research" category this year presumably would have been included under "other" or "applied mathematics" in earlier years. The counts of joint majors are in addition to the separate individual listings for mathematics, computer science or statistics majors since the total number of "bachelors degrees awarded by your department" was specifically asked for.

The 50% increase in the number of bachelors degrees in mathematics education since 1980 is encouraging. It should be noted that the questionnaire was sent to mathematics departments, per se, and in many universities and some public colleges, mathematics education students are handled separately by colleges or departments of education rather than by mathematics departments. Thus the figures cited are understood to be incomplete as counts of the total number of secondary education graduates in mathematics. However, the trend data should be meaningful since the counts over time are comparable.

The actual number of degrees in the mathematical sciences including joint majors but not including computer science or mathematics education as such has gone from 19,403 in 1974-75 to 12,154 in 1979-80 to 17,529 in 1984-85. The increase since 1979-80 was about 44%.

The totals of mathematics, statistics and computer science degrees may be compared with Department of Education figures for the July 1984 to June 1985 period which show 15,146 mathematics degrees (including 371 statistics majors) and 38,878 computer and information science majors. With some uncertainties as to how to classify some applied mathematics degrees and whether (all?) information science degrees would have been counted in the CBMS Survey, the figures appear to be generally consistent with Survey data.

Recent data from the 1986 AMS Survey which counts majors in school for the junior-senior years show a slight decrease in such majors in the mathematical sciences over the past year and a larger decrease in computer science majors over that period.

THE DISTRIBUTION OF BACHELORS DEGREES GRANTED IN 1984-85 AMONG VARIOUS TYPES OF DEPARTMENTS

Table 1-9 below gives the distribution of majors by type of department. It should be noted that "Mathematics Departments" is the catch-all category for universities or colleges which do not have separate statistics or computer science departments; such mathematics departments normally perform (part of) the functions of departments in those disciplines.

TABLE 1 - 9

NUMBERS OF BACHELORS DEGREES BY TYPE OF DEPARTMENT, JULY 1984-JUNE 1985

	Math Dept.	C.S. Dept.	Stat. Dept.	Total
Mathematics (general)	11,956	146	0	12,102
Applied Mathematics	1,215	0	0	1,215
Math. Education	2,567	0	0	2,567
Computer Science	8,646	20,416	45	29,107
Statistics	212	0	326	538
Operations Res.	302	0	10	312
Joint Mathematics & C.S.	2,519	565	0	3,084
Joint Math. & Statistics	102	0	19	121
Joint C.S. & Statistics	2	148	7	157
Total	27,521	21,275	407	49,203

There are several items in the table worthy of note. As expected, all mathematics education degrees are from mathematics departments. About 82% of joint mathematical and computer science majors are reported by mathematics departments. The development of computer science major programs within mathematics departments must be preceded by extensive course programs in computer science. Thus although the number of computer science sections taught in mathematics departments is almost the same as in computer science departments, we should not expect mathematics departments to produce as many computer science degrees as do computer science departments.

In Table 1-10, numbers of bachelors degrees in mathematics departments are shown by type of institution. For comparison purposes, the total FTE Faculty Size (Full-time plus 1/3 Part-time) is given in the bottom line. It gives partial support to the common belief that private colleges, with their attention to undergraduates, do turn out proportionately somewhat more bachelors degrees in the mathematical sciences. It is not clear whether the computer science category should be included in such comparisons. In any event, the non-existence of competing engineering and various specialty degree programs in private colleges presumably contributes to the observed differences. Note how applied mathematics degrees are concentrated in universities and public colleges and mathematics education degrees in the colleges. However as noted above, in some universities and public colleges, mathematics education degrees are the province of colleges or schools of education and thus are not included in the counts given. The higher incidence of separate computer science departments in universities presumably accounts for the smaller number of computer science degrees in university mathematics departments.

TABLE 1 - 10

1984-85 BACHELORS DEGREES FROM MATHEMATICS DEPARTMENTS
BY CATEGORY OF SCHOOL FOR VARIOUS DEGREE TYPES

	Univ.	Pu. 4-Yr.	Pr. 4-Yr.	Total
Mathematics (general)	3,467	4,277	4,212	11,956
Applied Mathematics	624	537	54	1,215
Mathematics Educ.	324	1,376	867	2,567
Computer Science	1,865	3,175	3,606	8,646
Statistics	115	97	0	212
Operations Research	259	43	0	302
Joint C.S. & Math	605	1,102	811	2,519
Joint Math. & Stat.	25	77	0	102
Joint C.S. & Stat.	0	2	0	2
Total	7,284	10,686	9,551	27,521
FTE Faculty Total Size (for comparison purposes)	5,681	8,866	5,664	

See Table 4-11 for a separate breakdown of degrees from computer science departments by category of institution. See the Introduction for a discussion indicating that the categories are not directly comparable to AMS Survey Group I, II & III; Group M; and Group B.

STATISTICS AS AN UNDERGRADUATE SUBJECT

The data generally cited elsewhere but organized below gives much information about undergraduate statistics. Enrollments in statistics in departments of the mathematical and computer sciences has grown rapidly over the past twenty-five years, at both elementary and advanced levels. We classify probability as a part of statistics for this purpose. Total enrollments in probability courses themselves are quite small and some include a probability and statistics designation. See Appendix B or E for Course titles.

TABLE 1 - 11

UNDERGRADUATE ENROLLMENTS IN STATISTICS OVER TIME
(in Thousands)

	1960	1965	1970	1975	1980	1985
Elem. Stat./Prob. Courses 45, 46	4	11	57	99	104	144
Adv. Stat./Prob. Courses 47-54	16	32	35	42	43	64
Total	20	43	92	141	147	208

We may conjecture on various reasons for the continuing impressive growth of undergraduate statistics enrollments:

(1) The increasing quantification of society, causing numerical data, its collection, use, analysis and interpretation to be much more widespread.
(2) The developing computer age which underlies much of (1) above.
(3) The increasing student choice of business as a major subject and the computerization and quantification of the whole business community resulting in statistics and probability and their applications becoming an integral part of the business curriculum.

The distribution of types of statistics courses among universities, public colleges and private colleges is revealed in Table 1-12. (See Appendix E for individual course enrollments).

TABLE 1 - 12

1985 STATISTICS COURSE ENROLLMENTS BY CATEGORY OF INSTITUTION
(in Thousands)

	Univ.	Pu. 4-Yr.	Pr. 4-Yr.	Total
Elem. Stat./Prob. (No Calc. prereq. 45, 46)	52	54	39	144
Math. Stat./Prob. (Calculus prereq. 47, 48)	17	14	9	39*
Other Stat. Courses (49-54)	20	4	1	25

*   Total from original data

The distribution of elementary courses is roughly proportional to the distribution of pre-calculus non-remedial courses in mathematics departments except that the public four-year college figure above is too low.   But as the courses get more specialized, the colleges show relatively low course enrollments. With the courses 47-54 lumped together the enrollments are roughly proportional to the numbers of statisticians on the three faculties (see Table 2-12).

The numbers of degrees in statistics has been reported by the Survey only for 1974-75, 1979-80, and 1984-85.

TABLE 1 - 13

NUMBERS OF STATISTICS UNDERGRADUATE DEGREES OVER TIME

1974 - 75	1979 - 80	1984 - 85
570	467	816

The figure 816 is from 538 reported as statistics majors, 121 as joint mathematics and statistics majors and 157 as joint computer science and statistics majors. Of these 278 joint majors only 26 were from statistics departments.   Thus, since in previous Surveys there was no place to list joint majors, it seems very likely that in earlier years almost all joint majors in statistics would have been listed only as mathematics or computer science majors.   Of the 538 statistics degrees, 326 were from statistics departments and 212 from mathematics departments.   In addition, mathematics departments produced 302 operations research degrees and statistics departments produced 10.   Statistics departments also produced 45 computer science majors.   The total number of degrees reported by statistics departments was 407 with mathematics departments reporting another 514 in statistics or operations research.   Thus with other joint statistics majors, there were 1,173 majors with a very large dose of statistics.   Unfortunately, the available data from the earlier surveys does not give us a basis for a comparison of these latter numbers over time.

For information on statisticians on the faculty, see Table 2-12.

REMEDIAL MATHEMATICS

Table 1- 14 below shows the enrollments in the four remedial courses since 1975 in the various categories of institutions.

TABLE 1 - 14

ENROLLMENTS OVER TIME IN REMEDIAL COURSES BY CATEGORY OF INSTITUTIONS
(in thousands)

	Arith.	Gen. Math.	Elem. Alg.	Inter. Alg.
Univ.				
1975	----	----	4	26
1980	2	4	13	44
1985	3	2	15	36
Pu. 4-Yr.				
1975	5	23	22	46
1980	11	37	54	48
1985	8	18	52	77
Pr. 4-Yr.				
1975	1	3	L*	9
1980	1	8	7	12
1985	4	11	8	17

*L means some but less than 500

With remedial mathematics courses playing an important role in many departments' instructional and faculty loads, a special one-page supplemental questionnaire on remediation was sent to all sampled mathematics departments. The response rate was noticeably less than the response rate from four-year college and university mathematics departments and thus the data below is not as reliable as the rest of the data. The reader is referred to other parts of this chapter and to Appendix E for additional enrollment and trend data on remedial mathematics.

Remedial mathematics was not explicity defined but in the four-year college and university questionnaires the courses listed as remedial (high school) were arithmetic, general math (basic skills), elementary algebra and intermediate algebra (high school). The data cited below are summaries from all responses considered together without regard to type of institution and without projecting by strata to the total population.

a) 19% of the academic units administering the remedial programs were outside the mathematics department.

b) 34% of the units handling remediation reported follow-up studies on success rates of students.

c) 35% of the faculty are full-time with 36% of the full-time faculty being tenured and another 30% on tenure track.

d) 18% of the combined full- and part-time faculty staffing the remedial program have doctorates with an additional 49% having master's degrees. Of the doctorates, 27% have their degrees in mathematics education and 19% have their degrees outside mathematics or mathematics education.

e) Course load credit practices varied from 65% giving credit in arithmetic to 90% in intermediate algebra (high school).

f) Credit-toward-graduation practices varied from 10% for arithmetic to 61% for intermediate algebra (high school). However, the question was worded, "Is credit toward graduation given", with "yes" and "no" boxes to check. Thus if credit were given only in some very special curricula or under special circumstances such as a student not having high school credit for the course, the "yes" box would presumably have been checked. It is known from other sources that, in many institutions, majors in mathematics, engineering or physical science receive no credit toward graduation for any course below calculus.

g) The percentage of all remedial sections taught by part-time faculty varied from 34% in intermediate algebra to about 45% in each of arithmetic, general mathematics (basic skills) and elementary algebra.

CHAPTER 2

FOUR-YEAR COLLEGE AND UNIVERSITY FACULTY CHARACTERISTICS

This chapter deals with characteristics of those faculty teaching mathematics, statistics and computer science. In Chapter 4, there is considerable additional information on the faculty teaching computer science. And Chapter 3 includes some data on instructional methods, on computer usage, and on teaching assistants. See Chapter 6 for two-year college faculty characteristics.

TERMINOLOGY USED FOR FOUR-YEAR COLLEGE AND UNIVERSITY FACULTY

The mathematics (departmental) faculty refers to all members of the nation's mathematics departments. The statistics (departmental) faculty refers to all members of statistics departments separate from mathematics departments. The mathematical sciences (departmental) faculty refers to the combined mathematics and statistics (departmental) faculties. The computer science (departmental) faculty refers to all members of separate computer science departments. Thus it is disjoint from the mathematical sciences departmental faculty. The total computer science faculty refers to the computer science departmental faculty together with all members of the mathematical sciences departmental faculty who taught at least one computer science course in their own department in fall 1985. Members are full- or part-time in this total faculty according as they taught computer science full- or part-time.

Any FTE (full-time equivalent) faculty size is computed as the size of the full-time faculty plus one-third the size of the part-time faculty. The references to the mathematical and computer science faculty in higher education refer to the combined two- and four-year college and university faculty.

37

HIGHLIGHTS

■ The full-time faculty in the mathematical and computer sciences grew by 21% in the period 1980-1985 and now numbers 22,194 while the part-time faculty grew by 46% to 9,189.

■ Since 1970, the FTE (full-time equivalent) faculty of all those teaching in the mathematical sciences in four-year colleges and universities increased by 6% while the course enrollments in the mathematical sciences increased by 41%. In the same period, the overall FTE faculty in the mathematical and computer sciences grew by 40% while course enrollments grew by 72%.

■ The total computer science faculty (i.e. teachers of computer science) is now 5,651 full-time (3,605 in computer science departments), 5,342 part-time (1,984 in computer science departments) for a total 7,432 FTE, up from an estimated 1,182 FTE in 1970.

■ The percentage of doctorates among the full-time faculty in the mathematical and computer sciences has decreased from 82% in 1975 to 73% in 1985 in the overall four-year college and university category.

■ The percentages of tenured faculty in the mathematical and computer sciences have decreased to 66%, 54% and 49% in the university, public four-year college and private four-year college categories while the non-tenured non-doctorate full-time faculty percentages have increased to 7%, 17% and 28% respectively.

■ The net outflow (outflow minus inflow) of the mathematical sciences faculty to industry, business and government was about 1/2% of the total faculty in 1984-1985.

■ Teaching load assignments generally are similar to those in 1970. Typical computer science and statistics faculty teaching assignments are less than those for mathematics faculty.

- Of all sections taught by full-time and part-time faculty, full-time professorial level (assistant to full) faculty teach about two-thirds, other full-time faculty teach about one-seventh, and part-time faculty teach the rest (almost one-fifth). GTA's teach about 20% of all sections in universities and under 10% in public four-year colleges.

FACULTY IN HIGHER EDUCATION

In Table 2-1 we give data on faculty size for all of higher education (from Department of Education publications: Digest of Educational Statistics and Projections of Educational Statistics).

In Table 2-2 we give the comparable data on faculty size for the mathematical and computer sciences combined (from the CBMS Survey). The totals for both tables refer to so-called "senior" faculty including those at the instructor or comparable level but not including graduate assistants.

TABLE 2 - 1

FULL- AND PART-TIME FACULTY AND FTE's IN HIGHER EDUCATION
(In Thousands)

	1970	1975	1980	1985
Full-time	369	440	458	456
Part-time	104	188	236	254
Full-time Equivalents	402	501	538	534
4-Yr. Coll. & Univ. FTE	333	397	420	409*
2-Yr. College FTE	68	104	118	124*

* These two numbers are estimates and are probably slightly too low and too high respectively.

TABLE 2 - 2

FULL-, PART-TIME AND FTE FACULTY IN THE MATHEMATICAL
AND COMPUTER SCIENCES IN HIGHER EDUCATION

	1970	1975	1980	1985
Full-time	21,922	22,807	23,927	28,471
Part-time	5,042	7,009	12,975	16,622
FTE	23,603	25,143	28,252	34,012
4-Yr. Coll. & Univ. FTE	17,986	18,062	20,409	25,257
2-Yr. College FTE	5,617	7,081	7,843	8,755

While the overall FTE faculty in higher education has increased 33% since 1970, the mathematics and computer science faculty has increased by 44%. However in the same period, the FTE enrollments in higher education have gone up 30% (see Graph 1-A) while enrollments in the mathematical and computer sciences have gone up 74% (see Graph 1-B). Thus over the period 1970-1985 there has been a serious deterioration in the relative size of the overall mathematical and computer science faculty.

FACULTY SIZE TRENDS IN THE MATHEMATICAL AND COMPUTER SCIENCES

The faculty size changes in various categories of institutions and types of departments are shown in Tables 2-3 and 2-4. The 1985 figures generally agree with AMS Survey figures when allowances are made for known differences in the composition of the different sets of categories used. See the Introduction for more discussion of this issue.

TABLE 2 - 3

DEPARTMENTAL FACULTY SIZES IN FOUR-YEAR COLLEGES AND UNIVERSITIES

	1970 Full	1970 Part	1975 Full	1975 Part	1980 Full	1980 Part	1985 Full	1985 Part
Universities								
Math. Dept.	6235	615	5405	699	5605	1038	5333	1044
Stat. Dept.	700	93	732	68	610	132	662	103
C.S. Dept.	688	300	987	133	1236	365	1448	491
Public Colleges								
Math. Dept.	6068	876	6160	1339	6264	2319	7754	3337
Stat. Dept.	N/A		N/A		N/A		78	15
C.S. Dept.	N/A		N/A		436	361	1554	862
Private Colleges								
Math. Dept.	3352	945	3579	1359	4153	2099	4762	2706
C.S. Dept.	N/A		N/A		N/A		603	631
TOTAL	17,043	2,829	16,863	3,598	18,304	6,314	22,194	9,189

The data above show a 21% increase in the overall full-time mathematical and computer science four-year college and university faculty and a 46% increase in the part-time faculty from fall 1980 to fall 1985. These faculty increases occurred when the mathematical and computer science student enrollments, Table 1-6A, were increasing by 20%. Thus the period from 1980 to 1985 has seen our faculty size slightly more than keep up with student enrollments.

In Table 2-4, we give the same counts summed different ways.

TABLE 2 - 4

MATHEMATICS, STATISTICS AND COMPUTER SCIENCE DEPARTMENTAL FACULTY SIZES IN FOUR-YEAR COLLEGES AND UNIVERSITIES

	1970	1975	1980	1985
Mathematics Depts.				
Full-time	15,655	15,144*	16,022	17,849
Part-time	2,436	3,397	5,456	7,087
FTE	16,467	16,276	17,841	20,211
Statistics Depts.				
Full-time	700	732	610*	740
Part-time	93	68	132	118
FTE	731	755	654	779
Computer Science Depts.				
Full-time	688	987	1,672	3,605
Part-time	300	133	726	1,984
FTE	788	1,031	1,914	4,266

* This may represent an undercount.

The departmental faculty size data in Table 2-4 show a rather slowly growing mathematics departmental faculty and a much more rapidly growing computer science departmental faculty over the past fifteen years. Since the mathematics departmental faculty teaches a considerable amount of computer science, it is perhaps even more appropriate to separate out the total computer science faculty and look at the trends over time. In Chapter 4, we have a count of the total computer science faculty (i.e. the faculty who taught computer science in Fall 1985), which shows 5,651 full-time and 5,342 part-time for an FTE total of 7,432. The total FTE faculty that taught mathematics is the difference between the total FTE mathematical and computer science faculty (from the last columns of Table 2-3) and the total FTE computer science faculty. Table 2-5 gives the breakdown of the total mathematical and computer science faculty into those who teach the mathematical sciences and those who teach computer science, along with course enrollments in the mathematical and computer sciences. While the enrollments per FTE faculty member in the total computer science faculty have stayed almost constant from 1970 to 1985, the corresponding ratio for the mathematical sciences teaching faculty has gone up dramatically. Over the past fifteen years, while the mathematical sciences faculty has been helping to create, and spin off, the computer science faculty, it has been seriously neglected in terms of its own growth.

TABLE 2 - 5

FTE FACULTY AND COURSE ENROLLMENTS

	1970	1985
FTE Mathematical Science Teaching Faculty	16,804	17,825
Mathematical Science Enrollments	1,296,000	1,827,000
Enrollments per FTE Faculty	77	102
Total FTE Computer Science Faculty	1,182*	7,432
Computer Science Enrollments	90,000	558,000
Enrollments per FTE Faculty	76	75

* The FTE computer science departmental count of 788 from Table 2-4 is the only count available from 1970. To be comparable to the 1985 figures, the count should include other teachers of computer science. We have arbitrarily assigned a 50% factor to 788 to get the 1,182 total. This total is consistent with the later enrollment-faculty ratio.

The faculty counts do not include any allowance for graduate teaching assistants. As stated above, the FTE mathematical science teaching faculty is used in Table 2-5 in a special sense: it is the complement of the total FTE computer science faculty. From Table 2-5 it follows that the FTE faculty of all those teaching mathematical science courses increased 6% from 1970 to 1985 while the course enrollments in the mathematical sciences increased 41%.

43

The data given here is for combined mathematical and computer sciences faculty and enrollments. The undergraduate course enrollments per FTE faculty member have stabilized and, in fact, dropped slightly since 1980. The trends since 1965 reflect the rapid faculty expansion in the sixties, followed by fairly stable total faculty numbers in the seventies, while enrollments were going up in both mathematics and computer science, and in the eighties, the faculty expansion in the public and private college sectors with more modest enrollment increases.

GRAPH 2 - A

COURSE ENROLLMENTS PER FTE TOTAL MATHEMATICS
AND COMPUTER SCIENCE FACULTY

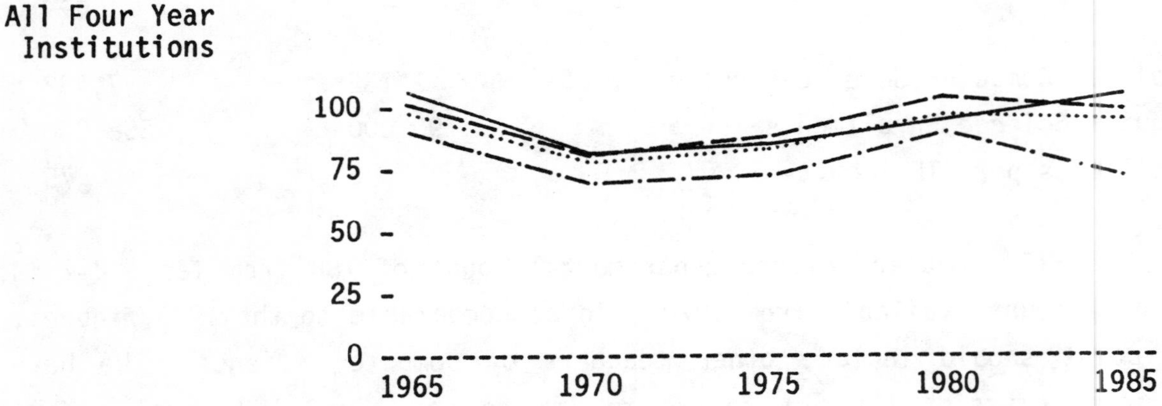

All Four Year
Institutions

	1965	1970	1975	1980	1985
Universities ————————	104	79	85	96	105
Public Colleges————	101	78	87	105	100
Private Colleges—·——·—	90	71	73	90	73
All Institutions ············	99	77	83	98	95

The data do not include either graduate teaching assistants on the one hand or graduate enrollments on the other.

44

TEACHING LOADS

In each Survey questionnaire since 1970 there has been a question concerning expected or typical teaching loads of faculty. Because of the increasing incidence of full-time faculty below the assistant professor level (see Tables 2-9 and 2-10 for current data on the percentages of sections taught by various components of the faculty), it was decided in this Survey to separately collect and report the data on teaching loads for such faculty (Table 2-8).

The percentages of departments in various categories reporting various expected loads over the past 15 years are given in Tables 2-6 through 2-8. There are several observations worth making.

(1) University statistics and computer science departments have consistently had noticeably lower loads (median 6 hours) as contrasted with university mathematics departments (median 7 hours).

(2) The public and private colleges have consistently had median loads at the 12-hour level except for public four-year college computer science departments which had a median 10-hour load in 1985.

(3) In all categories for which data from 1970 are available, the 1985 load patterns are quite similar to the 1970 patterns.

TABLE 2 - 6

EXPECTED OR TYPICAL CREDIT-HOUR TEACHING LOADS IN <u>MATHEMATICS</u> DEPARTMENTS
(PROFESSORIAL FULL-TIME FACULTY)

Percentage of Departments Indicating Given Load Per Semester or Quarter

	< 6	6	7-8	9-11	12	>12
Univ.						
1970	8%	40%	32%	13%	7%	--
1975	--	26%	39%	26%	10%	--
1980	10%	23%	29%	30%	9%	--
1985	11%	27%	36%	16%	10%	--
Pu. 4-Yr.						
1970	--	3%	5%	39%	35%	18%
1975	--	1%	5%	15%	57%	21%
1980	--	3%	6%	11%	59%	22%
1985	3%	4%	4%	19%	50%	20%
Pr. 4-Yr.						
1970	--	--	--	24%	60%	16%
1975	--	4%	2%	24%	56%	14%
1980	2%	3%	5%	24%	45%	22%
1985	--	--	6%	10%	64%	20%

The 1985 data refer to mathematics faculty teaching in the mathematical sciences, not in computer science. However, the computer science teaching loads in mathematics departments are quite similar. The 1970, 1975, and 1980 data presumably also refer primarily to professorial level faculty since that was the dominant faculty and only one percentage was recorded.

46

TABLE 2 - 7

EXPECTED OR TYPICAL CREDIT-HOUR TEACHING LOADS IN STATISTICS AND COMPUTER SCIENCE DEPARTMENTS
(PROFESSORIAL FULL-TIME FACULTY)

Percentage of Departments Indicating Given Load Per Semester or Quarter

	< 6	6	7-8	9-11	12	>12
Univ. Statistics						
1970	44%	28%	12%	16%	--	--
1975	17%	45%	11%	22%	5%	--
1980	9%	41%	34%	16%	--	--
1985	25%	54%	3%	17%	--	--
Univ. Comp. Sci.						
1970	17%	46%	27%	7%	3%	--
1975	14%	34%	19%	28%	5%	--
1980	24%	44%	8%	20%	4%	--
1985	25%	39%	25%	5%	5%	--
Pu. 4-Yr. Comp. Sci.						
1980	--	7%	--	23%	54%	15%
1985	6%	14%	17%	19%	34%	10%

It seems clear from Tables 2-6 and 2-7 that over the past five years, with the exception of private college mathematics departments, standard teaching loads are holding steady or dropping slightly. The data on private colleges seem inconsistent with the drop in course enrollments per FTE faculty member in that category, data given with Graph 2-A. But the data on recent trends in universities toward slightly lower expected loads also runs counter to the fifteen year rise in course enrollments per FTE faculty, data given with Graph 2-A. Perhaps the increase in the incidence of lecture sections in university departments and the hiring of more non-professorial faculty explain how expected professorial teaching loads can be kept low in the face of rising enrollments per FTE faculty member. Another factor affecting university faculty teaching loads is the graduate component. The Survey data do not specifically address this issue.

Since 1980 there has been a dramatic increase in the percentage of the faculty which is full-time non-doctorate and non-tenured (Table 2-12). A good many of such faculty would be expected to be in the non-professorial component of the faculty.

TABLE 2 - 8

1985 EXPECTED OR TYPICAL CREDIT-HOUR TEACHING LOADS
(NON-PROFESSORIAL FULL-TIME FACULTY)

Percentages of Departments Indicating Given Load Per Semester or Quarter

	< 6	6	7-8	9-11	12	>12
Math. Depts.						
Univ.	8%	9%	5%	20%	46%	12%
Pu. 4-Yr.	2%	3%	--	8%	59%	28%
Pr. 4-Yr.	--	--	9%	11%	62%	18%
Stat. Depts.						
Univ.	38%	7%	6%	42%	7%	--
Comp. Sci. Depts.						
Univ.	18%	11%	14%	41%	16%	--
Pu. 4-Yr.	3%	7%	4%	16%	58%	12%

The combination of low loads for a few and high loads for many suggests that there are two different types of full-time faculty below the professorial level:

(1) a fairly small number of research-type instructors chiefly in university departments and;

(2) a much larger number of faculty hired primarily as teachers with relatively high course loads.

TEACHING BY VARIOUS GROUPS OF FACULTY

This year, for the first time, the Survey reports on a more detailed analysis of teaching responsibilities by professorial level faculty (assistant to full), by other full-time faculty and by part-time faculty for the three categories of institutions and various types of departments. The results summarized in Tables 2-9 and 2-10 below show a rather consistent pattern: full-time professorial level faculty teach about 2/3 of the sections taught, other full-time faculty teach about 1/7 of the sections taught, and part-time faculty teach the remainder. The teaching of TA's was not included in these data. See the discussion following Table 2-10 and also Tables 3-4 to 3-6 for other data on teaching assistants. Table 2-9 refers to sections taught within mathematics departments only and Table 2-10 to sections taught in computer science and statistics departments.

TABLE 2 - 9

MATHEMATICS DEPARTMENT SECTIONS TAUGHT BY FULL- AND PART-TIME FACULTY
Rows sum to 100%

	Assistant to Full Professors	Other Full-Time	Part-Time
Mathematics Sections			
Univ. (n=12,185)	70%	14%	16%
Pu. 4-Yr. (n=21,489)	67%	14%	19%
Pr. 4-Yr. (n=11,727)	72%	10%	18%
Statistics Sections			
Univ. (n=759)	77%	10%	13%
Pu. 4-Yr. (n=1,912)	80%	10%	10%
Pr. 4-Yr. (n=1,531)	67%	10%	23%
Computer Science Sections			
Univ. (n=681)	78%	7%	15%
Pu. 4-Yr. (n=3,999)	64%	15%	21%
Pr. 4-Yr. (n=5,064)	64%	15%	21%

TABLE 2 - 10

COMPUTER SCIENCE AND STATISTICS DEPARTMENT SECTIONS TAUGHT BY
FULL- AND PART-TIME FACULTY
Rows sum to 100%

	Assistant to Full Professors	Other Full-Time	Part-Time
Comp. Sci. Depts.[1]			
Univ. (n=3,208)	63%	18%	19%
Pu. 4-Yr. (n=4,869)	68%	15%	17%
Pr. 4-Yr. (n=2,313)	54%	13%	33%
Stat. Depts.[1]			
Univ. & Pu. 4-Yr. (n=1,212)	83%	7%	10%

To get an estimate on the percentage of sections in universities or public colleges taught by graduate teaching assistants, we can compare the total number of sections reported in each of mathematics, statistics and computer science for questions 3 and 6D of the main questionnaire (Appendix B). The former gives total numbers of sections taught and the latter, as compiled, the numbers taught by full and part-time faculty (not GTA's) in each of the mathematics, statistics and computer science categories. For universities, this analysis shows the percentage of sections taught by GTA's for each of the three subject areas to be close to 20%. For public colleges, the overall percentage of sections taught by GTA's is less than 10%. Thus, to include the teaching of GTA's, the percentages of sections reported taught in universities in Tables 2-9 and 2-10 should be reduced by about 20% of the figures shown. In public colleges the percentages should be reduced by somewhat less than 10% of those figures. In private colleges there are a negligible number of

---

[1] The percentages shown are of all sections taught by the departments indicated. They include a small number of sections in mathematics or in the other of the two fields.

GTA's. The figures on numbers of sections taught by GTA's, by this analysis are generally consistent with the numbers, from Table 3-5, of GTA's in various categories, reported as teaching their own classes.

From the data from which the above tables are obtained, and Table 2-3, we can also find the average numbers of sections taught by part-time faculty in various types of departments. These numbers are given in Table 2-11.

TABLE 2 - 11

AVERAGE NUMBER OF SECTIONS TAUGHT BY A PART-TIME FACULTY MEMBER

Math. Dept.	Stat. Dept.	Comp. Sci. Dept.	All Depts.
1.54	1.01	1.11	1.44

Since about 5/6 of the total part-time faculty is in public or private colleges where the median expected load is close to 12 hours and most sections almost certainly are 3 hours per week, an estimate of a part-time faculty member as roughly equivalent to 1/3 FTE is reasonable.

DOCTORATES AMONG FULL-TIME MATHEMATICAL AND COMPUTER SCIENCE FACULTY

The trends over time in the percentages of doctorates among the combined full-time mathematical and computer sciences faculty are given by category of institution in Graph 2-B. We do not know how to explain the non-decrease in the public college percentage since 1980. Being counter to the overall trend, there could be a sampling abnormality in either year or a recording error in 1980. The decreases in the university and private college sectors are consistent with the large increases in total faculty (Table 2-3), and the large increases in the non-doctorate non-tenured faculty (Table 2-12). The overall percentage of doctorates among the total full-time faculty in 1985 was 73%.

51

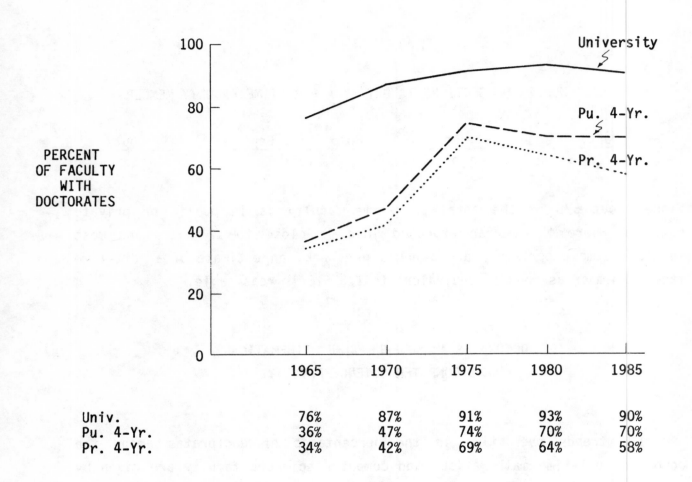

GRAPH 2 - B

PERCENTAGES OF FULL-TIME MATHEMATICAL AND
COMPUTER SCIENCES FACULTY HOLDING DOCTORATES

	1965	1970	1975	1980	1985
Univ.	76%	87%	91%	93%	90%
Pu. 4-Yr.	36%	47%	74%	70%	70%
Pr. 4-Yr.	34%	42%	69%	64%	58%

TENURE AND DOCTORAL STATUS OF THE FACULTY

In Table 2-12, we give 1975, 1980 and 1985 percentages of the total mathematical and computer sciences full-time faculty, with and without tenure and with and without doctorates. These data clearly show decreases in all categories of tenured doctorates and very marked increases in the

52

percentages of non-tenured non-doctorate faculty. From Table 2-9 concerning the distribution of the teaching of sections among professorial level faculty, other full-time faculty and part-time faculty, it follows that a good many of the non-tenured non-doctorate faculty reported in Table 2-12 for 1985 do have professorial status in both the public and private college categories.

TABLE 2 - 12

TENURE AND DOCTORAL STATUS OF TOTAL MATHEMATICAL AND COMPUTER SCIENCE FULL-TIME FACULTY FOR FALL 1975, 1980 AND 1985

	1975	1980	1985
Universities			
Tenured, PhD	67%	64%	63%
Tenured, non-PhD	5%	4%	3%
Non-tenured, PhD	26%	28%	27%
Non-tenured, non-PhD	2%	4%	7%
Public 4-Year			
Tenured, PhD	56%	52%[1]	51%
Tenured, non-PhD	18%	19%	13%
Non-tenured, PhD	20%	16%[1]	19%
Non-tenured, non-PhD	6%	13%	17%
Private 4-Year			
Tenured, PhD	45%	38%	35%
Tenured, non-PhD	25%	16%	14%
Non-tenured, PhD	24%	26%	23%
Non-tenured, non-PhD	6%	20%	28%
All Institutions			
Tenured, PhD	58%[2]	55%	51%
Tenured, non-PhD	14%[2]	12%	10%
Non-tenured, PhD	24%	23%	23%
Non-tenured, non-PhD	4%	10%	16%

(1) The figures given here from the 1980 report are slightly inconsistent with other numbers of faculty given on page 45 of that report showing 70% of the faculty with doctorates.

(2) The figures given in Table 3.13 on page 49 of the 1975-76 report for all institutions are inconsistent with the figures given there for various categories of institutions. The figures we use are based on a recalculation using faculty totals for the various categories given on page 48 of that report.

There are three factors, each of which would be expected to contribute to the modest reductions in percentages of total full-time faculty holding doctorates, from 82% in 1975 to 78% in 1980 to 73% in 1985, and the corresponding increases in the size of the non-tenured non-doctorate level faculty. First is the major increase in the teaching of remedial (high school level) mathematics between fall 1975 and fall 1980, from 141,000 course enrollments to 242,000. A doctorate is hardly a prerequisite for teaching courses at or below the level of second year high school algebra. Second is the major 21% expansion in the overall size of the full-time faculty between fall 1980 and fall 1985; there simply weren't enough PhD's available to maintain the percentage of doctoral holding faculty among all faculty. Third is the continuing major expansion of enrollments in computer science and, thus, in faculty teaching computer science. This occurs at a time when relatively few doctorates in computer science are being produced.

The percentages of tenured faculty and of doctorate-holding faculty are given in Table 2-13 by type of department and by category of institution. The overall percentages of the full-time mathematics departmental faculty and the full-time computer science departmental faculty that hold doctorates are 74% and 70% respectively (73% in the combined total faculty in the mathematical and computer sciences). Whereas 65% of the mathematics departmental faculty is tenured, only 42% of the computer science departmental faculty is tenured.

TABLE 2 - 13

TENURE AND DOCTORATE-HOLDING PERCENTAGES AMONG
FULL-TIME FACULTY IN 1985

	Tenured	Doctorate-Holding
Math. Depts.		
Univ.	71%	90%
Pu. 4-Yr.	69%	70%
Pr. 4-Yr.	51%	62%
Computer Sci. Depts.		
Univ.	49%	82%
Pu. 4-Yr.	40%	69%
Pr. 4-Yr.	31%	35%
Statistics Depts.		
	68%	97%

AGE AND SEX DISTRIBUTION OF FULL-TIME FACULTY

The age distributions of the full-time faculty in the mathematical and computer sciences for 1975 and 1985 are given in Table 2-14. The total faculty size in 1985 was 32% above that in 1975. In light of this increase in faculty size and assuming balanced attrition and new entrants, a 16% 40-44 age cohort in 1975 should be a 12% 50-54 age cohort in 1985. The figures given in Table 2-14 verify this observation for middle-level age groups.

At the younger age groups there will have been considerable attrition as well as new hirings over the 10 year span so that one does not expect the cohort to stay the same absolute size.

TABLE 2 - 14

AGE DISTRIBUTION OF FULL-TIME FACULTY IN THE
MATHEMATICAL AND COMPUTER SCIENCES

	1975	1985
< 30	10%	7%
30-34	22%	15%
35-39	22%	16%
40-44	16%	20%
45-49	11%	16%
50-54	9%	12%
55-59	5%	8%
60 or more	5%	6%

It is clear that the faculty is aging but not very rapidly. In the ten year span 1975-1985, the average faculty age has gone up from about 40.5 to over 43. Considering the big influx of new (younger) entrants, that seems about right.

The large increases in the part-time and in the non-tenured non-doctoral full-time components of the faculty, indicate that there is likely to be continuing turnover in the full-time faculty, producing, in the future, a more evenly spread out age distribution with a slowly increasing average age.

It should be noted that some individual university faculties, including some of the better research faculties, are reported to have aging problems. But the picture in-the-large does not look discouraging except for the drop in the under 40 populations.

The age distribution of the university mathematics faculty is almost identical to that of the overall mathematical sciences faculty in Table 2-15, differing by one or less in every age group percentage and having the average age of 44. Only 35% of the university faculty is under 40, in contrast to 55% in 1975 and 48% in 1980, a fact that may be of some concern. It is difficult to state what the norms should be. From Table 2-15 and from comparable figures in the university sector it is clear that the faculty has not been going on to retire at age 70 but rather has been retiring in the early to mid-sixties.

The men on the mathematical sciences faculty average about 44.5 years, four years older than the women. In computer science, the men average about 40.5 years of age, and the women 37. In Table 2-15 we give the age distributions of the separate departmental full-time faculties in the mathematical and computer sciences.

TABLE 2 - 15

AGE DISTRIBUTION OF MATHEMATICAL SCIENCE AND COMPUTER SCIENCE
DEPARTMENTAL FACULTIES

	< 30	30-34	35-39	40-44	45-49	50-54	55-59	60 or more
Math. Sci.	6%	14%	15%	19%	17%	14%	8%	7%
Comp. Sci.	13%	18%	20%	21%	13%	8%	5%	2%

MINORITIES

The percentages of minorities on the four-year college and university full-time faculty in each of the mathematical and computer sciences is given in the following table.

TABLE 2 - 16

MINORITIES IN THE FULL-TIME MATHEMATICAL AND COMPUTER SCIENCES DEPARTMENTAL FACULTIES

	Amer./Al.	Asian	Black	Hispanic
Math. Sci. Faculty	0.1%	7.1%	3.5%	3.4%
Computer Sci. Fac.	0.1%	11.7%	.3%	1.2%

In 1980 the Survey reported almost 3% of the faculty were Black. The current figures, compared with the 1975 Survey, do show a noticeable growth in the number of Asians on our faculties over the past 10 years, generally compatible with the growth (reported in other studies) of Asians among the graduate student population. In statistics departments, 22.5% of the faculty are Asian.

The Black faculty members are concentrated in the public college sector (6.9% of the total faculty there), presumably reflecting the faculty at historically Black institutions. In the university sector slightly less than 1% of the faculty is Black. The Hispanic mathematics faculty members are spread proportionately over all sectors of the four-year and university populations (except for statistics).

WOMEN ON THE FACULTY IN 1985

The percentage of the full-time departmental faculty who were female was 15% for mathematics, 10% for statistics, and 13% for computer science for an overall 14%, (the same percentage as that reported in 1980). As mentioned above, the average age of faculty women was about four years less than that of men.

Table 2-17 gives the percentages of faculty who are female by various types of departments and categories of institutions.

TABLE 2 - 17

PERCENTAGE OF FULL-TIME 1985 DEPARTMENTAL FACULTY WHO ARE FEMALE

- - - Mathematics Depts.- - -			- -Computer Science Depts.- -			Stat. Depts.
Univ.	Pu. 4-Yr.	Pr. 4-Yr.	Univ.	Pu. 4-Yr.	Pr. 4-Yr.	Combined
11%	19%	15%	11%	13%	23%	10%

DOCTORATE-HOLDING FACULTY

The 1985 age distribution of the doctorate holding mathematical and computer sciences faculties are available from two sources: (1) this Survey and (2) the biennial NSF publication, "Characteristics of Doctoral Scientists and Engineers in the United States". The gross figures are compatible although the details of the age distributions vary in the two studies. The NSF data is for academically employed doctoral scientists rather than for faculty, per se. But the two sets of data should be roughly comparable.

The total count of doctoral-holding mathematical scientists in this Survey is 13,025 compared to 13,027 in the NSF figures. The total count

59

of doctoral-holding members of the computer science departmental faculty is 2,537 in this Survey while the NSF figure for doctoral-holding computer specialists is 5,124, twice as much. The NSF figure presumably includes people in computer centers and on special research projects and presumably includes some faculty who are in mathematics or other departments but teach computer science. From the Survey's special computer science questionnaire, there are a total of 3,754 doctoral-holding full-time teachers of computer science and 2,231 doctoral-holding part-time teachers of computer science. Of the 2,231 part-time doctoral-holding faculty, 181 have degrees in computer science and 1,360 in mathematics and many are full-time in the reporting institution. Thus the NSF and Survey figures seem to represent comparable populations.

The age distribution of the doctoral-holding faculty from the Survey data and of academic scientists from NSF data are given in Table 2-18. It is not clear how the conflicting age patterns on the tails of the distributions should be reconciled. The current Survey data are quite consistent with past Survey data.

TABLE 2 - 18

AGE PATTERNS OF DOCTORAL FACULTY (SURVEY) AND ACADEMIC SCIENTISTS (NSF)

	< 30	30-39	40-49	50-59	60 or more
Mathematics					
Survey	5%	26%	40%	22%	7%
NSF Data	3%	25%	40%	22%	10%
Computer Science					
Survey	8%	37%	41%	11%	3%
NSF Data	2%	39%	40%	13%	6%

STATISTICS FACULTY

For all full-time faculty with <u>highest</u> degrees in statistics, Table 2-19 indicates their 1985 employment in mathematics, statistics and computer science departments.

TABLE 2 - 19

EMPLOYMENT OF STATISTICIANS IN VARIOUS TYPES OF DEPARTMENTS

	Univ.	Pu. 4-Yr.	Pr. 4-Yr.	Total
Mathematics Depts.	283	488	158	929
Statistics Depts.	572	38	---	610
Computer Science Depts.	20	6	29	55
Total	875	532	187	1594

Of these total full-time faculty, the numbers with doctorates are: universities, 814; public four-year colleges, 468; and private four-year colleges, 118; for a total of 1,400 in all university and four-year colleges. The Survey estimate on the total number of separate statistics departments in universities is 40 and in public colleges, 5, with none identified in private colleges. Because the numbers are so small in the public and private college sectors, it is quite probable that the sampling procedures used did not reveal the actual numbers. However the total faculty and PhD counts of statisticians should be fairly reliable. It is known (and is consistent with the data above) that many universities and some colleges which do not have separate statistics departments do have groups of statisticians on their mathematics faculties acting as subdepartments. For information on students in statistics see Tables 1-11 to 1-13.

FACULTY MOBILITY IN THE MATHEMATICAL SCIENCES

Data on faculty mobility from academic year 1984-85 to academic year 1985-86 are given in Table 2-20 for the total full-time mathematical sciences departmental faculty. Details are not given by category of institution because the overall numbers concerned are fairly small. In the 1975 and 1980 Survey reports, comparable data were given for the combined mathematical and computer sciences departmental faculty. This year, results for computer science departments are given separately in Chapter 4. The data this year do not show significant differences from the 1980 data except that the overall hiring of non-doctorate faculty from graduate schools shows an increase of about 60% from the 1980 figures and the "other" categories this year are relatively larger, though still small in absolute numbers. The "deaths and retirements" category for faculty outflow is about 1.1% of the total faculty, a small percentage. For the 1980 Survey, the combined mathematical and computer science faculty had a 0.9% death and retirement rate. Interesting figures are the Inflow/Outflow ratios with respect to non-academic employment which are 76/157 for doctorates and 116/33 for non-doctorates. The net loss to non-academia for doctorate faculty is only about a half percent of all doctoral faculty. The total faculty who switched departments (institutions) was about 600 doctorates and 175 non-doctorates.

TABLE 2 - 20

FULL-TIME MATHEMATICAL SCIENCES DEPARTMENTAL FACULTY MOBILITY
1984-85 to 1985-86

	Doctorates	Non-Doctorates
Faculty Inflow		
From Graduate School	362	463
From Post-Doctoral or Research Appointments	75	4
From Non-Academic Positions	76	116
From "Other"	45	80
Total Inflow	558	663
Faculty Outflow		
Deaths and Retirements	163	57
To Graduate School	27	90
To Non-Academic Positions	157	33
Otherwise Occupied	62	99
Total Outflow	409	279

It should be noted that these are one-year figures on mobility. They represent a one-year increase of 533 in total mathematical science departmental faculty which, with a 148 increase in computer science department faculty, is quite consistent with the reported overall full-time faculty growth from 18,304 in 1980 to 22,195 in 1985. The figures are also reasonably consistent with AMS Survey data which projected a total mathematical sciences faculty growth of 682 from fall 1983 to fall 1984.

The one-year Survey growth figures given above are from changes in existing departments. They do not include the creation or abolition of separate departments or institutions. However, chairpersons may well have reported only changes in that component of the faculty (still) in the department, if the department had been split.

NEW JOB OPENINGS FOR 1985-86

Question 10 on the main questionnaire tried to identify faculty openings and whether or not they were filled by people meeting the advertised qualifications. The reader is referred to Appendix B for the precise wording of this question. Because of spotty responses, the projections of the responses are statistically somewhat less reliable than those of most other questions. We present combined data for all four-year colleges and universities since there appeared to be only relatively minor differences between categories of institutions in most cases and the combined data is probably most reliable. Note from Table 2-21 that about three-fourths of all positions in mathematics and statistics and one-half of all positions in computer science were filled by people meeting advertised qualifications. Half the other openings were left unfilled.

TABLE 2 - 21

NEW JOB OPENINGS FOR 1985-86
PERCENTAGES OF OPENINGS FILLED IN VARIOUS WAYS

Qualif. Sought Number Sought	Math. Depts.			C.S. Depts.	Stat. Depts.
	Math. n=1502	C.S. n=598	Stat. n=149	n=784	n=68
Filled, Qualified	75%	44%	73%	53%	81%
Filled, Qualified, Part-time	5%	13%	5%	8%	4%
Filled, Unqualified	7%	15%	7%	15%	4%
Not-Filled	13%	28%	15%	24%	11%

Question 10, itself, avoids the issues of whether individual departments are realistic in terms of educational qualifications and salary levels for advertised positions. The results suggest that most departments are realistic.

CHAPTER 3

ADMINISTRATIVE ISSUES AND DEPARTMENTAL PHENOMENA

This chapter deals with changes in administrative structures affecting the mathematical and computer sciences, teaching loads, the uses of various instructional formats, computer use in instruction, the teaching functions and discipline sources of graduate teaching assistants, and issues considered important by department chairpersons.

HIGHLIGHTS

■ The creation of new computer science departments and the broadening of departmental duties and names to include computer science were frequent administrative changes.

■ In five major introductory courses, 41% of university students are taught in large lecture sections (over 80 students) whereas in private colleges only 2% are. About one-fifth of all students in these five courses are taught in sections of 40 to 80 students.

■ There is little required use of computers in college algebra or calculus or in any mathematics course other than numerical analysis or other computing related courses.

■ Since 1980 the number of graduate teaching assistants has been stable in university mathematics departments but has gone up markedly in statistics and computer science departments and in public college mathematics departments.

- About 95% of all graduate teaching assistants in mathematics, statistics or computer science are students in the same or related subjects.

- Salary levels and departmental support practices were widely regarded as major problems in many departments.

ADMINISTRATIVE STRUCTURES

The Survey questionnaire sought information on administrative changes in the period 1980-1985 affecting departments in the mathematical and computer sciences. Questions 2a and 2b (see Appendix B) referred to consolidations or divisions of departments. Table 3-1 gives administrative changes reported in university, public four-year college and private four-year college categories.

TABLE 3 - 1

1980-1985 ADMINISTRATIVE CHANGES AFFECTING MATHEMATICAL OR COMPUTER SCIENCES DEPARTMENTS

	Total Number Of Institutions	Consolidations	Divisions
Universities	157	12	22
Public 4-Year	427	80	62
Private 4-Year	839	158	75

Of the institutions reporting consolidations:

   (a)   about 40% involved formation of schools or divisions with several new mathematics and computer science departments included in these changes;

   (b)   about 35% were mathematics and computer science consolidations but in many instances these appeared to be simple expansions of mathematics departments and/or name changes;

   (c)   about 25% involved new departments such as mathematics and physics, computer science and electrical engineering, etc.

The "division of departments" reported are almost all accounted for as new computer science departments. In the three categories of institutions, new computer science departments from 1980 to 1985 were sepaarately calculated as 11, 59 and 102. Not all would have occurred as divisions of departments. In the university category a few other divisions into various mathematical science departments likely occurred.

As mentioned elsewhere, there are now projected to be separate computer science departments in 105 of the 157 universities, in 141 of the 427 public four-year colleges, and in 150 of the 839 private four-year colleges.

INSTRUCTIONAL FORMATS

The Survey sought information (Question 4) from all respondents as to the sizes or types of classes taught in selected introductory subjects. The specific question, a slight variant of that used in the 1980 Survey, asked for the numbers of students taught in:

   (1)   small classes (less than 40 students),

   (2)   large classes (between 40 and 80 students),

   (3)   lectures (over 80 students without recitation or quiz sections),

   (4)   lectures (over 80 students with recitation or quiz sections),

   (5)   self-paced instruction and

   (6)   other formats (See Appendix B).

The five subjects were College Algebra, Calculus (Math., Eng., Phys. Sci.), Calculus (Bio., Soc., Mgmt. Sci.), CS I (Computer Prog. I), and

Elementary Statistics. (The subjects were those used in the 1980 questionnaire except that "college algebra" replaced "finite math" since enrollments in college algebra were much higher than those in finite math).

To clarify the question, the various courses were listed with the identifying numbers used on the questionnaire form. The statistical analysis (projections) of the results were complicated by occasional incomplete or misleading answers to this particular question. However the overall results were generally consistent with those reported in 1980. In the 1975 Survey, a different type question was used and thus results from 1975 cannot be compared directly.

Generally, there were two major findings of which only the first is Vvident from Table 3-2:

(1) There are sharp differences in instructional formats between universities, public four-year colleges and private four-year colleges and

(2) Within any of the 3 categories of institutions, the reported differences in formats for the five subject areas studied were rather minor, particularly for the two calculus courses and computer programming. College algebra generally was taught in somewhat smaller sections and statistics in somewhat larger.

TABLE 3 - 2

PERCENTAGE OF STUDENTS TAUGHT IN VARIOUS FORMATS
IN FIVE STANDARD INTRODUCTORY COURSES

Class Format	University		Public 4-Year		Private 4-Year	
	1980	1985	1980	1985	1980	1985
<40	36	38	67	62	79	82
40 - 80	31	20	21	22	13	16
>80, no Quiz Sec.	10	12	2	5	1	-
>80, Quiz Sec.	21	29	9	10	7	2
Self-Paced or Other	1	1	-	1	-	-

The table shows that in universities there appears to be a trend away from large classes (40-80) toward lectures with quiz sections. In universities more than 40% of students in these five subject areas are taught in a large lecture format whereas in private colleges only 2% are.

It is also worthy of note that a negligible number of students (less than 1% in these five subject areas) are taught in "self-paced" or "other" modes; the standard formats totally predominate. The mathematical community is definitely not convinced of the efficacy of non-standard modes of instruction when it comes to course content needed for further work. The 1975 Survey showed that there was widespread experimentation with various alternative forms of instruction. It is clear from the 1980 and 1985 results that in the basic introductory courses the standard formats totally predominate. Similar turning away from various alternative modes or forms of instruction was evident in the two-year college category (see Chapters 5 and 6).

COMPUTER USE IN INSTRUCTION

All respondents were asked to indicate (Question 5) the number of sections in various courses in which the use of computers (micros/ minis/ mainframes) is required. A comparable question had not been asked in 1980. The responses were not of good statistical quality. Thus the results, listed as percentages of the total number of sections for the named courses, are summaries of all responses rather than projections. The results reveal relatively little obligatory computer use in mathematics courses except for those subjects closely identified with computing or computation. See Chapter 4 for a discussion of computer use in computer science courses and Chapter 5 for related two-year college phenomena.

TABLE 3 - 3

REQUIRED COMPUTER USE IN MATHEMATICS AND STATISTICS COURSES
AS PERCENTAGE OF ALL SECTIONS TAUGHT IN SELECTED SUBJECTS

Course and Number	Univ.	Pu. 4-Yr.	Pr. 4-Yr.	Total
College Algebra (5)	0%	0%	3%	0%
Calculus (15)	5%	8%	6%	7%
Diff. Equations (17)	15%	11%	13%	13%
Discrete Math. (18)	11%	28%	19%	18%
Linear Algebra (19)	15%	7%	23%	13%
Numerical Analysis (37)	91%	85%	82%	87%
Elementary Statistics (45)	29%	23%	43%	29%

The issue of the (required) use of calculators in mathematics or statistics courses in four-year colleges or universities was not pursued in the 1985 Survey.

GRADUATE TEACHING ASSISTANTS

In the 1980 Survey there were two questions about teaching assistants: one about the total number of teaching assistants, including the numbers who were graduate students in various types of departments (those reporting, other mathematical (computer?) science or not mathematical science) or, who were undergraduates. The second question dealt with the utilization of teaching assistants (teaching own classes, conducting quiz or recitation sections, paper grading, tutoring, other). There seemed to be ambiguity about the term "teaching assistant" if undergraduates were counted: e.g. "Are undergraduate paper graders teaching assistants?" In 1980, 50% of all mathematics teaching assistants in private colleges were tutors, whereas only 8% of teaching assistants in the university category were tutors, presumably indicating an uncertainty as to whether or not to count undergraduate students. From the last three columns of Table 3-4, it appears evident that some undergraduate paper graders or tutors were counted in various categories in 1980.

To clarify the terms in use, it was decided to request information for graduate teaching assistants only in the 1985 Survey. Thus the 1985 data are not directly comparable to the 1980 data, particularly in the private college category. According to the 1980 Survey report "more than one-fifth" of all teaching assistants reported were undergraduates. See Table 3-5 for a comparison of counts of teaching assistants or graduate teaching assistants reported in 1980 and 1985. In Table 3-5, the private college mathematics category clearly reveals a count of many undergraduates classed as teaching assistants in 1980. For 1985, with only GTA's included, the number of teaching assistants was much lower. For other phenomena, see the discussion following Table 3-5.

In Table 3-4 are given the 1985 percentage distribution of graduate teaching assistants by principal teaching function by category of department.

The reader should note, as pointed out in the introduction to this report and in Chapter 2, that the university and public college categories are not identical, or directly comparable with, the AMS Survey Groups I, II, and III and M. For technical reasons, the Department of Education lists from which the Survey sampling was drawn produces a set of institutions for the university category which effectively replaces a number of large public universities in AMS Groups I, II, and III with smaller private universities. These larger universities then appear in the public four-year college category. Thus the Survey totals on GTA's in universities would be expected to be somewhat lower than in AMS Group I, II, and III data and the public four-year college totals would be expected to be somewhat higher than in AMS Group M data. That is the case. But the overall totals are not inconsistent.

TABLE 3 - 4

PRINCIPAL TEACHING FUNCTIONS OF GRADUATE TEACHING ASSISTANTS
1985
(Rows sum to 100%)

Type of Department:	Teaching Own Class	Conducting Quiz/Recit. Sections	Paper Grading	Tutoring	Other
University					
Math. (n=5038)	47%	40%	8%	4%	1%
Stat. (n=711)	24%	52%	14%	6%	4%
C.S. (n=1746)	36%	26%	23%	11%	4%
Public 4-Year					
Math. (n=2077)	44%	41%	9%	6%	0%
Stat. (n=85)	29%	15%	56%	0%	0%
C.S. (n=530)	23%	15%	35%	24%	3%
Private 4-Year					
Math. (n=111)	60%	34%	3%	3%	0%
C.S. (n=30)	40%	20%	40%	0%	0%

The high incidences of "paper grading" and/or "tutoring" functions in statistics and computer science departments probably reflect the different nature of homework or projects in those subject areas as compared to mathematics. They presumably reflect both (1) the handling of data and/or computers, requiring different types of activities and knowledge than grading freshman mathematics papers and (2) some different patterns of instruction including a higher percentage of lecture sections.

The following table of reported numbers of graduate teaching assistants (teaching assistants for 1980) with their principal teaching function is perhaps even more revealing than would be corresponding percentages. The numbers for which data are given for 1980 are extracted from the totals and percentages reported in 1980, since the actual numbers for 1980 are not available.

TABLE 3 - 5

NUMBER OF TEACHING ASSISTANTS FOR 1980 AND
GRADUATE TEACHING ASSISTANTS FOR 1985
BY PRINCIPAL TEACHING FUNCTION

	Teaching Own Class	Conducting Quiz/Recit. Sections	Paper Grading	Tutoring	Other
University					
Mathematics Depts.					
1980	2745	1592	604	439	55
1985	2368	2015	403	202	50
Statistics Depts.					
1980	44	229	153	120	0
1985	171	369	100	43	28
Computer Science Depts.					
1980	329	381	653	471	0
1985	629	453	402	192	70
Public 4-Year					
Mathematics Depts.					
1980	445	230	230	414	230
1985	913	852	187	125	0
Computer Science Depts.					
1980	23	0	51	15	0
1985	122	80	185	127	16
Private 4-Year					
Mathematics Depts.					
1980	81	219	277	577	0
1985	67	38	3	3	0

The universally higher numbers in the "paper grading" and "tutoring" functions for 1980 over 1985 (except for Public 4-Year Computer Science Departments with almost twice as many departments in 1985) strongly suggest that many undergraduates assigned to these functions were counted as teaching assistants in 1980. The sum of the actual numbers in columns 1 and 2 for university mathematics departments shows a small gain from 1980 to 1985 in teaching assistants actually teaching rather than a small loss superficially suggested by gross data. The impressive five year gains in columns (1) and (2) for most public college and university categories indicate a substantially broader use of teaching assistants for teaching, consistent with the generally sizeable increases in part-time and non-professorial full-time faculties in these categories (compare with the figures in Table 2-4).

By comparing the 1980 and 1985 data and questions, it seems clear that some private colleges as well as public four-year colleges use some undergraduates for teaching functions.

It would be nice to have fairly reliable estimates of the percentage changes in the numbers of graduate teaching assistants in the various categories of departments for which we have data on the number of teaching assistants in 1980. Based on estimates from the "over 20%" figure of undergraduates among the 1980 teaching assistants and from an analysis of the principal teaching functions of teaching assistants in 1980 and 1985, it seems clear that:

(1)   the number of graduate teaching assistants in university mathematics departments in 1985 was substantially the same as that in 1980;

(2)   the number of graduate teaching assistants in university computer science departments who actually performed teaching functions increased by about 50% from 1980 to 1985, and;

(3)   the number of mathematics department graduate teaching assistants in public four-year colleges who actually performed teaching functions more than doubled from 1980 to 1985.

In other categories the 1980 figures were sufficiently small and the procedures too uncertain to make meaningful estimates of percentage increases from 1980 to 1985.

With respect to the public four-year college computer science category, it should be noted that the number of departments went up 100% from 1980 to 1985.

WHAT DO GRADUATE TEACHING ASSISTANTS STUDY?

In the 1980 Survey report, it was stated that in university mathematics departments more than 20% of teaching assistants were not mathematics graduate students. The exact figure was not given, nor was there a breakdown into undergraduate or graduate students in another statement that "more than 20%" were in "other departments". Table 3-6 below gives information for graduate teaching assistants in 1985. Overall, almost all (92%) of the graduate teaching assistants are students in the department for which they teach and half of the rest are students in other mathematics or computer science departments.

TABLE 3 - 6

DEPARTMENTS IN WHICH GRADUATE TEACHING ASSISTANTS STUDY
1985

	Number	Percentage In Own Department	Percentage In Other Mathematics Or Computer Sci. Dept.
University			
Mathematics	5,038	91%	6%
Statistics	711	91%	2%
Computer Science	1,746	98%	2%
Public 4-Year			
Mathematics	2,077	86%	1%
Statistics	85	100%	0%
Computer Science	530	100%	0%
Private 4-Year			
Mathematics	111	92%	8%
Computer Science	30	100%	0%
Total	10,328		

The widely reported earlier use by state university mathematics departments of graduate teaching assistants who were students in engineering or other disciplines seems to have largely ended.

Since this Survey dealt with undergraduate phenomena including, of course, teaching assistants, but not with graduate education, per se, there was no attempt to identify numbers of graduate research assistants or associates. There was also, perhaps regrettably, no attempt to identify citizenship status of graduate teaching assistants. AMS, NSF and other studies address parts of this latter issue.

DEPARTMENTAL CONCERNS

As a new initiative to give a statistical base for possible new studies on the status of the profession in academe, the Survey included two lists of questions, one on professional activities of faculty and how they affect faculty advancement and/or salary decisions and the second on problems of the mid-80's as seen by department chairpersons. Each question had a scale of 0 to 5 with zero representing no importance and 5 representing major importance. The results were tabulated for all categories of institutions for both mathematics and computer science departments and for university statistics departments. The numbers shown in the remaining tables in this chapter are (1) the percentage of all departments giving a 4 or 5 response for the particular question and (2), in parentheses, the percentage giving a 0 or 1 response for the same question. The percentage giving a 2 or 3 response can be found by subtracting the sum of the two percentages given from 100. The difference in the two numbers given is a measure of the preponderance of departmental attitudes on the subject. Note that high percentages do not measure the intensity of feeling, as such, but rather the breadth of concern.

IMPORTANCE OF PROFESSIONAL ACTIVITIES

Table 3-7 gives the results of the questionnaire on the importance of various professional activities in faculty advancement and/or salary decisions by category of institution. The results confirm conventional wisdom, university departments value published research and colleges, particularly private colleges, value teaching performance. Service to the department or institution is much more commonly important to colleges than to universities. In universities, mathematics departments and statistics departments have remarkably similar priorities.

Generally the computer science department responses on professional activities were quite similar to the mathematics department responses. They are given separately in Chapter 4.

TABLE 3 - 7

IMPORTANCE OF PROFESSIONAL ACTIVITIES IN ADVANCEMENT AND/OR SALARY DECISIONS

	- - - Mathematics - - - -			Stat.
	Univ.	Pu. 4-Yr.	Pr. 4-Yr.	Univ.
Classroom Teaching Performance	70 (3)	81 (2)	96 (4)	71 (6)
Published Research	96 (0)	70 (10)	26 (39)	100 (0)
Service to Department and/or University (College)	31 (5)	63 (5)	66 (0)	31 (11)
Talks at Profess. Mtgs.	42 (5)	49 (11)	13 (28)	25 (11)
Activities in Profess. Societies and/or Pub. Service	22 (8)	45 (4)	33 (9)	31 (6)
Supervision of Grad. Students	34 (7)	21 (32)	-------	81 (0)
Undergraduate/Grad. Advising	9 (22)	24 (20)	39 (12)	21 (21)
Years of Service	1 (52)	34 (29)	46 (16)	15 (47)
Expository and/or Pop. Articles	22 (13)	37 (14)	14 (40)	14 (19)
Textbook Writing	9 (35)	17 (35)	11 (58)	12 (50)

PROBLEMS OF THE MID-80'S

In the 23 questions on problems of the mid-80's there were several questions where the responses stood out significantly. The results for these questions are given separately in Table 3-8A. The remaining results are given in three tables, (1) those dealing with student issues, Table 3-8B, (2) those dealing with faculty issues, Table 3-8C and those dealing with support issues, Table 3-8D. We give the results in approximate order of decreasing importance as seen by departmental chairpersons. The concerns not commonly considered as major problems are almost as interesting as those considered important. Generally, those concerns identified as major problems are those which need addressing by the community. As to be expected, for some concerns there are wide variations reported among departments in the various types of universities or colleges and between departments in the mathematical and computer sciences. The responses for computer science are summarized separately in Chapter 4 in Tables 4-18 and 4-19A to D of this report. The heavy emphasis on salary and support issues reported in Table 3-8A means there is continuing pressure for upward salary adjustments and that there should be continuing pressure for better departmental support services. Clearly the larger community should be concerned with departmental support practices.

As in Table 3-7, the percentage of departments identifying the concerns as of major (minor) importance is given in Tables 3-8A to D.

TABLE 3 - 8A

MAJOR PROBLEMS

	Univ.	Pu. 4-Yr.	Pr. 4-Yr.	Stat. Univ.
Salary Levels/Patterns	66 (6)	69 (2)	60 (0)	64 (8)
Departmental Support Services (Travel, Secret. etc.)	61 (13)	62 (10)	36 (23)	70 (0)
Research Funding	71 (8)	45 (17)	17 (63)	51 (14)
Maintaining Faculty Vitality	54 (13)	54 (5)	41 (21)	48 (18)

TABLE 3 - 8B

STUDENT ISSUES

	Univ.	Pu. 4-Yr.	Pr. 4-Yr.	Stat. Univ.
Lack of Quality of Undergraduate Majors	38 (15)	62 (6)	39 (7)	31 (9)
Lack of Quantity of Undergraduate Majors	39 (18)	54 (20)	42 (9)	22 (21)
Lack of Quality of Department Graduate Students	50 (2)	44 (21)	------*	56 (14)
Lack of Quantity of Department Graduate Students	52 (14)	53 (24)	------*	55 (20)
Remediation	39 (28)	66 (5)	45 (17)	0 (42)
Class Size	52 (12)	39 (21)	21 (32)	60 (6)

*   Since relatively few of the departments in this category have graduate programs, the responses are not given.

TABLE 3 - 8C

OTHER FACULTY ISSUES

| | Mathematics | | | Stat. |
	Univ.	Pu. 4-Yr.	Pr. 4-Yr.	Univ.
Teach. Load of Full-Time Fac.	44 (22)	59 (17)	59 (9)	40 (27)
The Need to Use Temporary Fac.	42 (18)	44 (28)	42 (31)	35 (32)
Promotion-Tenure Process Above Departmental Level	24 (47)	39 (26)	15 (29)	36 (22)
Advancing Age of Tenured Fac.	29 (24)	25 (31)	14 (39)	21 (41)
Lack of Experienced Senior Fac.	11 (55)	14 (48)	15 (51)	33 (52)
Losing Full-Time Faculty to Industry/Government	15 (48)	10 (64)	9 (65)	51 (34)

TABLE 3 - 8D

OTHER SUPPORT ISSUES

| | Mathematics | | | Stat. |
	Univ.	Pu. 4-Yr.	Pr. 4-Yr.	Univ.
Upgrading/Maint. of Computer Facilities	34 (21)	42 (29)	46 (33)	48 (9)
Office/Lab Facilities	45 (23)	30 (29)	19 (37)	50 (23)
Computer Facilities (Classroom)	38 (18)	37 (31)	40 (23)	39 (18)
Classroom Lab Facilities	41 (16)	22 (25)	26 (33)	29 (20)
Computer Facilities (Fac. Use)	31 (23)	33 (32)	25 (30)	39 (18)
Networking Facilities	26 (35)	30 (35)	16 (46)	27 (27)
Library: Holdings, Access, etc.	20 (46)	25 (35)	10 (43)	16 (40)

CHAPTER 4

COMPUTER SCIENCE IN FOUR-YEAR COLLEGES AND UNIVERSITIES

In this year's Survey serious attempts were made to get more information concerning the status of computer science in undergraduate instruction. The titles of the Survey and questionnaire were changed to reflect the Survey's concern with undergraduate programs in the mathematical sciences and in the computer sciences. A special supplemental one page computer science questionnaire (see Appendix D) was sent to "those departments which offer undergraduate programs (not necessarily degree programs) in computer science."

COMPUTER SCIENCE REFERENCES

Specific references to various aspects of computing and computer science will also be found in Tables: 1-1, 1-6A, 1-6B, 1-8, 1-9, 1-10, 2-3, 2-4, 2-5, 2-7, 2-8, 2-9, 2-10, 2-11, 2-13, 2-15, 2-16, 2-17, 2-18, 2-19, 2-21, 3-3, 3-5, 3-6, 5-2, 5-3, 5-4, and 5-5.

The reader should refer to the beginning of Chapter 2 for explanations of faculty terms used in this report.

HIGHLIGHTS IN 1985

■ Two-thirds of all universities, one-third of all public four-year colleges, and more than one-sixth of private four-year colleges have separate computer science departments. In the public four-year college category the number is five-thirds that for 1980.

■ There were 5,651 members of the full-time total computer science faculty of whom 3,605 were in computer science departments. There were 5,342 part-time computer science faculty of whom 1,984 were in computer science departments.

■ Of the 3,754 doctorates who teach computer science full-time, 1,291 have their degrees in computer science and 1,555 in mathematics. Of the 2,231 doctorates who teach computer science part-time, 181 have their degrees in computer science whereas 1,369 have their degrees in mathematics.

■ Half of all part-time computer science faculty teach full-time in the same institution, almost a third are employed outside education and a tenth are not employed full-time anywhere.

■ Half (49%) of all computer science sections are taught in mathematics departments, the other 51% in computer science departments.

■ In a substantial number of institutions, some computer science is taught outside mathematics and computer science departments, chiefly in business, engineering and education academic units.

■ Total reported enrollments in computer science have climbed from 107,000 in 1975 to 321,000 in 1980 to 558,000 in 1985.

■ There were 29,107 computer science undergraduate degrees in fiscal year 1984-85, with 8,646 of these in mathematics departments. In addition there were 3,084 joint majors with mathematics. The number of computer science degrees reported in the 1980 Survey for fiscal year 1979-80 was 8,917.

■ About two-thirds of all institutions require calculus for computer science majors, one-half require linear or matrix algebra and more than two-fifths require discrete mathematics.

■ The most common problems reported by computer science departments are salary levels and patterns, departmental support services, the need to use temporary faculty, and the upgrading and maintenance of computer facilities.

NUMBERS OF SEPARATE COMPUTER SCIENCE DEPARTMENTS

In all, 155 of the special computer science questionnaires were returned with the following overall distribution

	Univ.	Pu. 4-Yr.	Pr. 4-Yr.
By Math. Depts.	13	42	36
By Comp. Sci. Depts.	35	22	7

reflecting the fact that in the universities there are many separate computer science departments whereas in the colleges some computer science is taught in many mathematics departments. These numbers are not identical with the numbers of such departments that returned the main questionnaire.

Based on estimates from reports from all institutions responding to any aspect of the questionnaire, the numbers of computer science departments in the various categories of institutions are given in Table 4-1 along with comparable data from the 1980 report.

TABLE 4 - 1

NUMBERS OF SEPARATE COMPUTER SCIENCE DEPARTMENTS
With Numbers of Institutions

	1980	1985
Univ.	94 of 160	105 of 157
Pu. 4-Yr.	85 of 407	141 of 427
Pr. 4-Yr.	48 of 830	150 of 839

COMPUTER SCIENCE FACULTY IN FALL 1985

Using data from both the special computer science questionnaire and the main questionnaire, we can identify many characteristics of those who taught computer science in 1985. We have numbers and various characteristics for the computer science departmental faculty and for the total computer science faculty, both full and part-time. The departmental faculty numbers are obtained from computer science departments on the main questionnaire. The total or overall faculty numbers are obtained from the special computer science questionnaire. The numbers for the faculty teaching computer science but not in computer science departments are obtained by subtracting the former from the latter. See Appendices B and D for copies of the questionnaires.

It is very important to note the implied definitions of the full-time and part-time components of the total computer science faculty. In Tables 4-2A through 4-6B, full-time (or part-time) refers to faculty teaching computer science full-time (or part-time). It does not refer to full-time (or part-time) faculty members at the institution. Table 4-5 makes it clear that about half of all part-time computer science faculty are, in fact, full-time at the same institution. In Table 4-8, there is a different analysis of computer science teaching phenomena wherein, for example, a full-time mathematics department faculty member teaching computer science part-time in the mathematics department would be classified as a full-time faculty member. It is worthwhile to compare the faculty and teaching divisions in Tables 4-2A to 4-6B with those in Table 4-8.

We begin by giving the numbers of the full-time and the part-time total computer science faculty in Tables 4-2A and 4-2B.

TABLE 4 - 2A

FULL-TIME TOTAL COMPUTER SCIENCE FACULTY

	Univ.	Pu. 4-Yr.	Pr. 4-Yr.	Total
C.S. Depts.	1,448	1,554	603	3,605
Other Depts.	91	853	1,102	2,046
Total	1,539	2,407	1,705	5,651

TABLE 4 - 2B

PART-TIME TOTAL COMPUTER SCIENCE FACULTY

	Univ.	Pu. 4-Yr.	Pr. 4-Yr.	Total
C.S. Depts.	491	862	631	1,984
Other Depts.	178	1,454	1,726	3,358
Total	669	2,316	2,357	5,342

The figures above are not surprising in light of the distribution of computer science departments as shown in Table 4-1. The part-time faculty will in many instances represent faculty in other departments at the same institution. See Table 4-5 for sources of part-time faculty. The fact that the number of part-time faculty is almost as large as the number of full-time faculty is of interest.

It should also be noted that in universities and colleges without computer science departments it would be expected that most, or in some cases all, of those who teach computer science would be part-time and generally borrowed from other departments.

FIELDS OF HIGHEST DEGREE FOR COMPUTER SCIENCE FACULTY

In Tables 4-3A and 4-3B are given (1) the numbers of full-time and part-time computer science faculty with highest degrees in various categories and (2), in parentheses, the percentages of those counted who have the doctoral degree. Thus, in Table 4-3A, 83% of the 796 full-time computer science faculty in universities with highest degree in computer science have doctoral degrees in computer science. Of course, the numbers of faculty with highest degree in various areas can be read independently of the percentages of doctorates.

Tables 4-4A and 4-4B below give the numbers of doctorates on the total computer science faculty with degrees in various areas. The numbers in the "other" categories may seem large but "other" includes the various physical and social sciences.

TABLE 4 - 3A

FULL-TIME TOTAL COMPUTER SCIENCE FACULTY BY FIELD OF HIGHEST DEGREE

The Parenthetical Percentages Show Those With Doctorates

Field of Highest Degree	Univ.	Pu. 4-Yr.	Pr. 4-Yr.	Total
Comp. Sci.	796 (83%)	990 (51%)	627 (20%)	2413 (54%)
Math.	388 (91%)	899 (83%)	670 (68%)	1957 (80%)
Engin.	131 (85%)	89 (87%)	37 (22%)	257 (76%)
Educ.	45 (69%)	56 (75%)	114 (68%)	215 (70%)
Stat.	32 (75%)	32 (0%)	21 (0%)	85 (29%)
Other	147 (88%)	341 (81%)	236 (56%)	724 (74%)
Total	1539 (85%)	2407 (68%)	1705 (47%)	5651 (67%)

From the last column in Table 4-3A we compute that 43% of the full-time faculty teaching computer science have their highest degree in computer science. From Table 4-4A, we may note that only 34% of the doctorates teaching computer science have their doctorates in that field.

TABLE 4 - 3B

PART-TIME TOTAL COMPUTER SCIENCE FACULTY BY FIELD OF HIGHEST DEGREE

The Parenthetical Percentages Show Those With Doctorates

Field of Highest Degree	Univ.	Pu. 4-Yr.	Pr. 4-Yr.	Total
Comp. Sci.	319 (25%)	820 (10%)	472 (4%)	1,611 (11%)
Math.	133 (65%)	845 (60%)	1,251 (62%)	2,229 (61%)
Engin.	68 (57%)	117 (32%)	122 (48%)	307 (41%)
Educ.	18(100%)	88 (67%)	21(100%)	127 (77%)
Stat.	23 (48%)	42 (36%)	68 (68%)	133 (55%)
Other	108 (65%)	404 (16%)	423 (57%)	935 (40%)
Total	669 (45%)	2,316 (33%)	2,357 (49%)	5,342 (41%)

The data in Tables 4-3A and 4-3B strongly support the evidence from Table 4-5 that a substantial part (50%) of the part-time total faculty in computer science is full-time faculty in the same institution and from Table 4-8 that it has a large component which is full-time mathematics faculty teaching computer science courses within the mathematics department.

DOCTORATE-HOLDING COMPUTER SCIENCE FACULTY

From the numbers and percentages in Tables 4-3A and B we can get a detailed analysis by their fields of degrees of those doctorates who teach computer science. Tables 4-4A and B below give these counts summed both ways. As in Tables 4-3A and B the counts are for all faculty teaching computer science.

TABLE 4 - 4A

DOCTORATES ON FULL-TIME TOTAL COMPUTER SCIENCE FACULTY

Field of Doctorate	Univ.	Pu. 4-Yr.	Pr. 4-Yr.	Total
Comp. Sci.	661	505	125	1,291
Math.	353	746	456	1,555
Engin.	111	77	8	196
Educ.	31	42	78	151
Stat.	24	0	0	24
Other	129	276	132	537
Total	1,309	1,646	799	3,754

There are two noteworthy observations from Table 4-4A.

(1) Of the doctorates who teach computer science full-time in universities, slightly more than half have their degrees in computer science and more than half of the rest have their degrees in mathematics.

(2) Of the doctorates who teach computer science full-time in the public and private colleges, almost half have their degrees in mathematics and more than half of the rest have their degrees in computer science.

TABLE 4 - 4B

DOCTORATES ON <u>PART-TIME</u> TOTAL COMPUTER SCIENCE FACULTY

Field of Doctorate	Univ.	Pu. 4-Yr.	Pr. 4-Yr.	Total
Comp. Sci.	80	82	19	181
Math.	86	507	776	1,369
Engin.	39	37	59	135
Educ.	18	59	21	98
Stat.	11	15	46	72
Other	70	65	241	376
Total	304	765	1,162	2,231

Almost two-thirds of the doctorates who teach computer science part-time in the public and private four-year colleges have their degrees in mathematics. A large number of part-time faculty with doctorates, particularly in the private college category, presumably are faculty from the same institution. Many of those with degrees in mathematics will be in mathematics departments as such, since mathematics departments in the college sectors teach a great deal of computer science. Note Table 4-5 where it is shown that 59% of all part-time computer science faculty in the private college sector are employed full-time in the same institution.

SOURCES OF PART-TIME COMPUTER SCIENCE FACULTY

The sources of part-time computer science faculty in terms of their full-time employment is given in Table 4-5. Each column adds to 100%.

TABLE 4 - 5

SOURCES OF REGULAR EMPLOYMENT OF PART-TIME TOTAL COMPUTER SCIENCE FACULTY
Columns sum to 100%

Employed Full-Time at:	Univ.	Pu. 4-Yr.	Pr. 4-Yr.	All
Own Institution	52%	42%	59%	50%
Other Univ. or College	7%	2%	5%	4%
High School	1%	3%	5%	4%
Outside Education	23%	41%	23%	31%
Not Employed Full-time Anywhere	17%	12%	8%	11%

It seems reasonably clear that part-time computer science faculty members are selected from whatever resources are available. Many private colleges are in small towns where the source of part-time faculty would be the same institution. Public colleges are perhaps more likely to be in or near larger centers where non-academic personnel are available. The data tend to support this analysis.

About half of the part-time faculty in computer science are employed full-time in the same institution, with more than three-tenths employed full-time outside academia and more than one-tenth not employed full-time anywhere. Some retired persons or faculty spouses employed to teach part-time may be in this last category.

THE BROADER COMPETENCE OF COMPUTER SCIENCE FACULTY

A question was designed to find how broadly competent the computer science faculty was judged to be: specifically, what percentages of the computer science faculty teach only lower-level courses or only specialty courses. It should be expected that much of the part-time faculty would be in such categories. From the responses recorded in Table 4-6B below it would appear that most chairpersons reported limits in qualifications in one or the other but not both "lower level" and "specialty" course categories.

TABLE 4 - 6A

PERCENTAGES OF FULL-TIME TOTAL COMPUTER SCIENCE FACULTY TEACHING ONLY LOWER LEVEL OR SPECIALTY COURSES

	Univ.	Pu. 4-Yr.	Pr. 4-Yr.	Total
Lower Level Courses	11%	18%	31%	20%
Specialty Courses	13%	18%	17%	16%

TABLE 4 - 6B

PERCENTAGES OF PART-TIME TOTAL COMPUTER SCIENCE FACULTY TEACHING ONLY LOWER LEVEL OR SPECIALTY COURSES

	Univ.	Pu. 4-Yr.	Pr. 4-Yr.	Total
Lower Level Courses	43%	58%	42%	49%
Specialty Courses	21%	11%	10%	12%

From Tables 4-6A and B we conclude that perhaps one-third of the full-time and three-fifths of the part-time computer science faculty teach only lower level or specialty courses.

FACULTY MOBILITY

Data on faculty mobility from academic year 1984-85 to academic year 1985-86 is available for the national faculty in computer science departments. Separate data for computer science faculty within mathematics departments is not available - the figures given in Table 2-20 are for faculty mobility in mathematics and statistics departments, including those who teach computer science there.

The data show an increase for the one year of 60 doctorate faculty and 88 non-doctorate faculty in computer science departments. From the nature of the question (#9 on the main questionnaire) it is likely that figures from departments newly created for 1985-86 are not included. Thus the total size of the national computer science departmental faculty should have increased somewhat more. It is interesting that the outflow/ inflow ratios to/from non-academic employment are 52/32 for doctorates and 70/48 for non-doctorates. The ratios for mathematics and/or statistics departments are 157/76 and 33/116. At the doctorate level the ratios are not dissimilar. There were also approximately 120 doctorates and 40 non-doctorates who went from one computer science department (school) to another.

TABLE 4 - 7

MOBILITY OF THE FULL-TIME COMPUTER SCIENCE DEPARTMENTAL FACULTY
1984-85 To 1985-86

	Doctorates	Non-Doctorates
Faculty Inflow		
From Graduate School	91	165
From Post-Doctoral or Research Appts.	21	0
From Non-Academic Positions	32	48
From Other Sources	6	0
Total Inflow	150	213
Faculty Outflow		
Died or Retired	5	21
Returned to Graduate School	23	34
To Non-Academic Positions	52	70
To Other Status	10	0
Total Outflow	90	125

Data on the field of study of either doctorates or non-doctorates is not available in this mobility study. Presumably the doctorates (and the non-doctorates as well) who are going back to graduate school are seeking training in computer science, per se.

The net outflow to non-academic positions was about 1.2% of the total departmental faculty.

WHERE AND BY WHOM IS COMPUTER SCIENCE TAUGHT?

The number of sections of computer science taught by various components of the nation's four-year college and university faculty is shown in Table 4-8 below. The total number of sections taught in mathematics departments, 9,744, is just under the total number taught in computer science departments, 10,102.

The definitions of "full-time and part-time faculty" are not the same as those used in Tables 4-2A and B and 4-3A and B and elsewhere in this chapter. The data for Table 4-8 came from the main questionnaire and part-time would refer to part-time in the department reporting. Thus mathematics department chairpersons would report a full-time departmental faculty member as "full-time" even though he/she taught only one or two computer science sections.

TABLE 4 - 8

PERCENTAGE OF SECTIONS OF COMPUTER SCIENCE TAUGHT
(Columns sum to 100%)

	Univ.	Pu. 4-Yr.	Pr. 4-Yr.	Total
Math. Depts.				
By Full-time Faculty	15%	36%	56%	39%
By Part-time Faculty	3%	10%	15%	10%
C.S. Depts.				
By Full-time Faculty	66%	45%	19%	40%
By Part-time Faculty	16%	9%	10%	11%

From the 36% and 10% figures for public colleges and the 56% and 15% figures for private colleges it follows that in both public and private four-year college mathematics departments almost four-fifths of the computer science taught there is taught by full-time mathematics department faculty members. The data above seems generally consistent with the data in Tables 4-2A and B to 4-4A and B. In part, it corroborates the preponderance of mathematically trained faculty among the teachers of computer science.

The percentage of all computer science sections taught in computer science departments ranges from 82% in universities, to 54% in public four-year colleges to 29% in private four-year colleges. These data agree reasonably well with the current ratios of numbers of computer science departments to numbers of institutions, Table 4-1, with the caveat that institutions with separate computer science departments would be expected to teach relatively more computer science than would those without computer science departments.

OTHER UNITS TEACHING UNDERGRADUATE COMPUTER SCIENCE COURSES

Departments were asked to identify units other than mathematics or computer science departments within the institution which taught computer science courses. The responses are summarized in the following table.

TABLE 4 - 9

OTHER UNITS TEACHING SOME UNDERGRADUATE COMPUTER SCIENCE COURSES
Percentages Of All Institutions Responding

	Univ.	Pu. 4-Yr.	Pr. 4-Yr.
Business	56%	53%	22%
Engineering	47%	22%	15%
Education	20%	27%	9%
Other Natural Science	17%	6%	15%
Computer Center	8%	10%	2%
Social Science	4%	10%	4%
Humanities	4%	2%	0%
Library	0%	5%	0%

The figures reported do not seem surprising. Some forms of computer science are taught rather widely in the institutions. Since elementary data processing is not listed as a computer science course, per se (See Appendix A or E), it seems likely that data processing in some form may account for much of the high incidence of teaching in the "business" category.

COMPUTER SCIENCE COURSE ENROLLMENTS

Since computer science as a subject has developed only over the past quarter century as the computer age has gone from a few very expensive mainframes to minis and micros and hand calculators, there have, of course, been massive changes in student enrollments. Indeed, the nature of computer science and of specific course content continues to change with the changing technology. It was only in the 1970 Survey (after the curriculum guidelines of ACM-68 were issued) that a detailed listing of computer science courses (more than 2 at any level) was used by the Survey. It was only with this year's Survey that a name distinction was made between the mathematical and computer sciences and that a separate chapter on computer science was introduced in the report.

Table 4-10 gives the trend in enrollments in computer science. The course numbers refer to courses listed in Appendix E.

TABLE 4 - 10

TRENDS IN COMPUTER SCIENCE COURSE ENROLLMENTS BY LEVEL
(in Thousands)

	1960	1965	1970	1975	1980	1985
Lower (55-61)	-	2	64	63	206	350
Middle (62-65)	-	12	12	19	35	66
Upper (66-92)	9	8	30	31	80	142
Total	9	22	106	113	321	558

The computing or computer science courses for 1960 and 1965 were listed along with mathematics courses. The titles "Programming for Digital Computers" and "Other Computer Science Mathematics" suggest subject matter now identified chiefly with elementary and middle level courses.

The distribution of enrollments by level for 1970 to 1980 are "best estimates" from specific course enrollments given in the Survey reports. Only with the current Survey were the lower, middle and upper level designations used.

Computer science, along with mathematics and statistics, has a major service component for other disciplines. The fact that 63% of the 1985 course load is at the elementary level supports this view. However, the very large number of majors in computer science, Table 4-11, means that unlike mathematics and statistics, a sizable part of the upper level enrollment is for those within the discipline.

COMPUTER SCIENCE STUDENTS

As noted in Chapter 1, Table 1-9, there were 29,107 computer science undergraduate degrees reported in 1984-85 plus another 3,084 joint majors with mathematics and 157 with statistics. Of the (single) computer science majors 20,416 were from computer science departments, 8,646 were from mathematics departments and 45 were from statistic departments. Of the joint majors with mathematics, 2,519 were from mathematics departments and 565 from computer science departments. Of the joint majors with statistics, all but nine were from computer science departments. As shown in Table 1-8, the reported numbers of computer science majors went from 3,636 in 1974-75 to 8,917 in 1979-80 to 29,107 in 1984-85.

The division of the computer science majors among universities, public four-year colleges and private four-year colleges is given in Table 4-11.

TABLE 4 - 11

1984-85 COMPUTER SCIENCE UNDERGRADUATE DEGREES BY CATEGORY
OF INSTITUTION AND DEPARTMENT
(Does not include joint majors)

Department	Univ.	Pu. 4-Yr.	Pr. 4-Yr.	Total
Comp. Sci.	9,122	8,335	2,959	20,416
Mathematics	1,865	3,175	3,606	8,646
Statistics	45	0	0	45
Total	11,032	11,510	6,565	29,107

The 2,519 joint mathematics-computer science majors from mathematics departments were distributed as follows: 605 in universities, 1,102 in public four-year colleges and 811 in private colleges. Of the 565 joint mathematics-computer science majors from computer science departments, 136 were in universities, 169 in public and 260 in private four-year colleges.

MATHEMATICS AND STATISTICS COURSES TAKEN BY
COMPUTER SCIENCE MAJORS

The special computer science questionnaire sought information on (1) the total number of mathematics and statistics semester or quarter courses (at the calculus level or above) <u>normally</u> taken by computer science majors and (2) the mathematics and statistics courses required of computer science majors. The average (mean) numbers of mathematics and statistics courses normally taken by computer science majors are shown below in Table 4-12. Thus computer science majors take very little more mathematics and statistics courses than do engineering majors.

Robert M. Aiken, Chair of the Education Board of the ACM, who reviewed this report, expressed some surprise at the data in Tables 4-12 and 4-13 and their implications. He states, "My experience in consulting with a number of programs and participating in computer science accreditation efforts leads me to believe that computer science majors take a minimum of two (mathematics and statistics) courses beyond the freshman-sophomore level." But he suggests that the fact that courses such as discrete mathematics (discrete structures) and numerical analysis are frequently taught within computer science departments and thus may be classified as computer science courses in this Survey may help explain the apparent discrepancy of his experience with Survey data.

TABLE 4 - 12

NUMBER OF SEMESTER OR QUARTER COURSES IN MATHEMATICS OR
STATISTICS NORMALLY TAKEN BY COMPUTER SCIENCE MAJORS

Number of Courses Taken in	Univ.	Pu. 4-Yr.	Pr. 4-Yr.
Math/Statistics	5.4	4.3	4.5

The table below lists the percentages of schools in the university and college categories which require various mathematics and statistics courses for computer science majors. The courses are listed in approximate decreasing order of frequency of being required. All courses for which at least 10% of departments in any category of institution require the course are listed. The course numbers are those identifying the courses in the Survey questionnaire (Appendix B or E).

TABLE 4 - 13

PERCENTAGES OF INSTITUTIONS REQUIRING THE GIVEN MATHEMATICS OR
STATISTICS COURSE FOR COMPUTER SCIENCE MAJORS

		Univ.	Pu. 4-Yr.	Pr. 4-Yr.
15	Calculus (Math, Phys. Sci., Eng.)*	88%	61%	69%
19	Linear Alg. & Matrix Theory	65%	44%	43%
18	Discrete Mathematics	48%	34%	47%
37	Numerical Analysis	27%	24%	24%
47	Math. Statistics (Calc. prereq.)	29%	15%	13%
27	Discrete Structures	21%	13%	17%
17	Differential Equations	22%	9%	6%
22	Combinatorics	13%	5%	0%
45	Elem. Stat. (no Calc. prereq.)	6%	13%	12%
46	Prob. & Stat. (no Calc. prereq.)	3%	2%	12%
48	Probability (Calc. prereq.)	12%	5%	11%
50	Applied Statistical Analysis	5%	10%	0%

*   The questionnaire does not reveal explicitly whether one, two or more
semesters (quarters) of calculus are required.

The results rather clearly support the view that undergraduate
computer science has evolved (or is evolving) into a discipline quite
distinct from mathematics.   Only about 70% of institutions require
computer science majors to take calculus, about 60% to take discrete
mathematics or discrete structures, only about 50% to take linear
algebra/matrix theory, about 50% to take some statistics course and 25% to
take numerical analysis.   Differential equations and combinatorics are
required of only a small percentage of majors.

Furthermore, the fact that the average computer science major takes
five or fewer semester (or quarter) mathematics or statistics courses
suggests that only a fairly small percentage of computer science majors

opt for (or are advised to take) more than core courses in freshman-sophomore mathematics. (But see the comments preceding Table 4-12).

TYPES OF COMPUTER SCIENCE DEGREES

Schools (departments) teaching computer science were asked to identify the type of degree, if any, offered in computer science. Some institutions have several types of degrees. Because the questionnaire was directed to departments offering computer science, per se, it is likely that many business-oriented data processing programs were simply not included as respondents.

TABLE 4 - 14

PERCENTAGES OF SCHOOLS WHICH OFFER COMPUTER SCIENCE
HAVING VARIOUS TYPES OF DEGREE PROGRAMS
(Columns Do Not Sum to 100%)

Type of Comp. Sci. Degree	Univ.	Pu. 4-Yr.	Pr. 4-Yr.
None	17%	22%	18%
Science	79%	64%	77%
Business	9%	14%	28%
Engineering	18%	5%	5%
Education	1%	2%	0%
Other	0%	6%	2%

STUDENT USE OF MICROS OR MINIS/MAINFRAMES

In computer science courses with programming projects, mini-computers and mainframes were used much more widely than micro-computers. The use reported is given in Table 4-15.

TABLE 4 - 15

USE OF TYPES OF COMPUTERS IN COMPUTER SCIENCE PROGRAMMING PROJECTS
Percentage of Students Enrolled at Given Levels

	Univ.	Pu. 4-Yr.	Pr. 4-Yr.
Lower Level C.S. Courses			
Micros	35%	40%	39%
Minis/Mainframes	65%	60%	61%
Middle or Upper Level C.S. Courses			
Micros	21%	26%	30%
Minis/Mainframes	79%	74%	70%

Thus about three-eighths of lower level students and one-fourth of middle or upper level students used micros in programming projects in computer science, the rest used minis or mainframes.

CONTROL OF WORK STATIONS

Data on the control of student work stations used in computer science courses are given in Table 4-16.

TABLE 4 - 16

PERCENTAGE OF TEACHING DEPARTMENTS HAVING CONTROL OF
STUDENT WORK STATIONS

	Univ.	Pu. 4-Yr.	Pr. 4-Yr.
Micros	53%	39%	61%
Minis/Mainframes	21%	28%	51%

Clearly departments teaching computer science in private colleges are much more likely to have control of student work stations, particularly for minis or mainframes. As expected, the teaching departments are more likely to control micro work stations than those for larger computers.

STUDENT ENROLLMENTS PER WORK STATION

For students taking computer science courses and using the computer in Fall 1985 we have the following pattern of work station availability.

TABLE 4 - 17

ENROLLED AND ACTIVE COMPUTER SCIENCE STUDENTS PER WORK STATION
Percentage of Departments by Category
Columns sum to 100%

No. of Students Per Work Station	Univ.	Pu. 4-Yr.	Pr. 4-Yr.
0-5	12%	24%	18%
6-10	33%	29%	60%
11-15	38%	15%	19%
16-20	9%	17%	0%
20 or more	8%	15%	3%

Thus in terms of the number of students sharing a work station, the private four-year colleges are, on the average, noticeably better off than the universities or public colleges.

DEPARTMENTAL CONCERNS IN COMPUTER SCIENCE DEPARTMENTS

The Survey included two lists of questions, one on professional activities of faculty and how they affect faculty advancement and/or salary decisions and the second on problems of the mid-80's as seen by department chairpersons. Each question had a scale of 0 to 5 with zero representing no importance and 5 representing major importance. The results tabulated here are in the university, public four-year college and private four-year college categories for computer science departments. The numbers shown in the following tables in this chapter are (1) the projected percentage of all departments giving a 4 or 5 response for the particular question and (2), in parentheses, the projected percentage giving a 0 or 1 response for the same question. Obviously the percentage giving a 2 or 3 response can be found by subtracting the sum of the two percentages given from 100. The difference of the two numbers given is a measure of the preponderance of departmental attitudes regarding the issue as important.

The responses of departments in the mathematical sciences are given at the end of Chapter 3 in similarly designed tables. To assist the reader in comparing the two sets of data, the grouping and order of listing of issues for the mathematical sciences and computer science departments are the same. The issues are listed in approximate decreasing order of importance as viewed by departments in mathematics and statistics. Thus the grouping and order for computer science departments may seen unnatural.

TABLE 4 - 18

IMPORTANCE OF PROFESSIONAL ACTIVITIES IN FACULTY ADVANCEMENT AND/OR SALARY DECISIONS

	Computer Science Departments		
	Univ.	Pu. 4-Yr.	Pr. 4-Yr.
Classroom Teaching Performance	48 (1)	90 (0)	67 (0)
Published Research	96 (0)	60 (5)	25 (33)
Service to Department and/or University (College)	37 (5)	54 (0)	46 (0)
Giving Talks at Profess. Mtgs.	39 (8)	43 (11)	9 (49)
Professional Activities in Profess. Societies and/or Pub. Service	36 (0)	21 (5)	9 (16)
Supervision of Graduate Students	40 (10)	47 (16)	-------
Undergraduate/Graduate Advising	5 (31)	31 (12)	21 (33)
Years of Service	5 (54)	42 (22)	63 (0)
Expository and/or Popular Articles	10 (26)	21 (14)	9 (16)
Textbook Writing	27 (25)	14 (35)	4 (66)

Classroom teaching performance is relatively more important in college departments than in university departments whereas published research is much more important in university departments. Professional activities including talks and textbook writing are of considerable importance in university departments and of little importance in college departments whereas years of service are generally important only in college departments.

Generally the computer science departmental responses on professional activities were quite similar to the mathematics department responses given separately in Table 3-7.

PROBLEMS OF THE MID-EIGHTIES

In the 23 questions on problems of the mid-80's there were several concerns where the responses stood out significantly. The results for four concerns are given separately in Table 4-19A. The remaining results are given in three tables, (1) those dealing with student issues, Table 4-19B, (2) those dealing with faculty issues, Table 4-19C and (3) those dealing with support issues, Table 4-19D. The concerns not commonly considered as major problems are almost as interesting as those considered important. Note that high percentages do not measure the intensity of feeling, as such, but rather the breadth of concern. Generally, those concerns identified as major problems are those which need addressing by the community. As is to be expected, for some concerns wide variations were reported among departments in the various types of universities and colleges and between departments in the mathematical and computer sciences. The responses for mathematics and statistics departments are summarized separately in Tables 3-8A to D.

The broad emphasis on salary and support issues identified in Table 4-19A means there is continuing pressure for upward salary adjustments and that there should be continuing pressure for better departmental support services. Clearly the larger community should be concerned with departmental support practices.

As in Table 4-18, the percentages of departments identifying the concerns as of major (minor) importance are given in Tables 4-19A to D.

TABLE 4 - 19A

MAJOR PROBLEMS

	Computer Science Departments		
	Univ.	Pu. 4-Yr.	Pr. 4-Yr.
Salary Levels/Patterns	61 (1)	89 (6)	58 (0)
Departmental Support Services (Travel, Secretarial, etc.)	81 (1)	60 (3)	54 (14)
Research Funding	60 (14)	73 (14)	58 (26)
Maintaining Faculty Vitality	32 (24)	38 (3)	25 (66)

TABLE 4 - 19B

STUDENT ISSUES

	Computer Science Departments		
	Univ.	Pu. 4-Yr.	Pr. 4-Yr.
Lack of Quality of Undergraduate Majors	14 (46)	28 (25)	37 (16)
Lack of Quantity of Undergraduate Majors	5 (80)	9 (57)	16 (34)
Lack of Quality of Department Graduate Students	29 (18)	12 (38)	------*
Lack of Quantity of Department Graduate Students	18 (43)	14 (56)	------*
Remediation	4 (46)	5 (48)	26 (37)
Class Size	49 (22)	40 (20)	30 (66)

* Since relatively few of the departments in this category have graduate programs, the responses are not given.

TABLE 4 - 19C

OTHER FACULTY ISSUES

	Computer Science Departments		
	Univ.	Pu. 4-Yr.	Pr. 4-Yr.
Teaching Load of Full-Time Faculty	43 (13)	39 (19)	51 (33)
The Need to Use Temporary Faculty	58 (13)	44 (19)	51 (0)
Promotion-Tenure Process Above			
Departmental Level	42 (20)	63 (20)	33 (59)
Advancing Age of Tenured Faculty	5 (60)	3 (51)	4 (86)
Lack of Experienced Senior Faculty	52 (24)	54 (10)	63 (33)
Losing Full-Time Faculty to			
Industry/Government	32 (20)	38 (21)	41 (49)

TABLE 4 - 19D

OTHER SUPPORT ISSUES

	Computer Science Departments		
	Univ.	Pu. 4-Yr.	Pr. 4-Yr.
Upgrading/Maint. of Computer			
Facilities	59 (21)	60 (0)	33 (30)
Office/Lab Facilities	54 (0)	54 (12)	4 (49)
Computer Facilities (Classroom)	33 (17)	50 (0)	21 (21)
Classroom Lab Facilities	44 (5)	40 (12)	4 (70)
Computer Facilities (Faculty Use)	30 (32)	59 (14)	21 (79)
Networking Facilities	33 (41)	49 (24)	41 (59)
Library: Holdings, Access, etc.	15 (34)	17 (28)	21 (0)

The issue of maintaining faculty vitality (see Tables 3-8A and 4-19A) was a major problem in mathematics departments but not in computer science departments. The quality and quantity of undergraduate majors and of graduate students were minor problems in computer science but important problems in the mathematical sciences. It is interesting that losing faculty to industry or government was not considered a major concern for most computer science departments. Generally, the responses were consistent with commonly perceived faculty age and supply and demand phenomena in computer science. There were major differences in responses on several support issues between university and public college departments of computer science on the one hand and private college departments on the other. Generally private college departments are much better satisfied with computer access and availability, class size and facilities.

CHAPTER 5

MATHEMATICAL SCIENCE OFFERINGS, ENROLLMENTS, AND INSTRUCTIONAL PRACTICES IN TWO-YEAR COLLEGES

This chapter reports estimated national enrollments in two-year college (tyc) mathematical science courses for fall 1985. The data are compared and contrasted with results of previous CBMS surveys of 1966, 1970, 1975, and 1980 and with general enrollment trends in two-year colleges. For information on the sampling procedure used in this survey, see the Introduction and Appendix A.

HIGHLIGHTS
(1980-1985)

■ Mathematical science enrollments remained almost unchanged, decreasing by 1%.

■ Overall tyc enrollments decreased by 2%.
* Part-time students continued to account for nearly two-thirds of all tyc students.
* Nearly two-thirds of tyc associate degrees are now in occupational programs.

■ Mathematical science faculty increased by 12%. Full-time and part-time sectors each increased by 12%.

■ Courses showing increases were as follows:
* Statistics increased by 29%.
* Calculus increased by 13%.

* Remedial course enrollments reached 482,000, increasing by 9% since 1980. They now account for 47% of all tyc mathematical science enrollments and two-thirds of all remedial enrollments in higher education.
* Other precalculus increased by 4%.
* Computing increased by 3%.

■ Courses showing decreases were as follows:
* Technical mathematics decreased by 56%.
* Business mathematics decreased by 42%.
* Mathematics for liberal arts decreased by 42%. Enrollments in this course are now below 1966 levels.

■ Access to computers increased and the impact of computers and calculators on mathematics teaching increased.

■ Mathematics labs continued to grow in popularity and now can be found in 82% of two-year colleges.

■ Self-paced instruction decreased sharply in the period 1980-1985.

AN OVERVIEW OF TWO-YEAR COLLEGES:
IS THE BOOM OVER?

During the 60's and 70's, no other sector of higher education grew as rapidly as did two-year colleges. In the 60's their enrollments tripled; in the 70's they doubled. But in the 80's two-year college enrollment growth stopped; the period 1980-1985 showed an actual decrease. In 1960 two-year colleges accounted for only one-sixth of all undergraduate enrollments in mathematics. Today, the figure is nearly one-third.

Explosive growth of such proportions was accompanied by changes in programs, student populations, and faculty populations. In the early 60's, most two-year colleges had a liberal arts orientation, serving as feeders for four-year colleges. By the mid-60's, program emphases had undergone considerable change. A host of new programs in occupational areas were introduced: data processing, dental hygiene, electronics, practical nursing, automotive mechanics, accounting, bricklaying, carpentry, and police and fire science, to name a few. Today, less than half of two-year college students are enrolled in college transfer programs. The growing majority of students are enrolled in occupational programs, and two-thirds of associate degrees are in occupational programs.

Most of the students of the 60's were 18- and 19-year old high school graduates, planning to transfer to four-year colleges. Most of them were single, white, male, and attending on a full-time basis. Today, two-thirds of students are over 21, one-third are married, some lack high school diplomas, one-fourth are minority students, and more than one-half are women. Nearly two-thirds of these students are attending on a part-time basis, and one-half start their studies after age 21. Many of these students require training in remedial mathematics (arithmetic, high school geometry, elementary and intermediate algebra, and general mathematics). The growth of remedial courses has been dramatic; today they account for nearly half of all two-year college mathematics enrollments. By way of contrast, calculus enrollments now account for only 10% of enrollments, down from 12% fifteen years ago but up slightly since 1980.

Faculty populations have also changed since 1960. Then nearly two-thirds of full-time faculty had previously taught in high schools. Many of them entered two-year colleges expecting to move up to teach calculus-level courses. In a short time, they found themselves teaching courses in arithmetic. Since then, economic pressures have resulted in a sharp swing toward the use of part-time faculty. In the mid-60's, full-timers outnumbered part-timers by two to one; today, part-timers outnumber full-timers. Another aspect of the economic times is the phenomenon of overload teaching. At present, 43% of all full-time faculty in mathematics are teaching overloads, most for extra pay.

Self-paced instruction appeared in a variety of forms in the 60's and 70's: CAI, audio tutorial, television, modules, PSI, and film. With the current decrease in class sizes, we note a sharp decrease in their popularity.

Additional details on trends in course offerings, faculty populations, and changes in two-year college teaching environments are given in the following pages.

TRENDS IN OVERALL TWO-YEAR COLLEGE ENROLLMENTS, 1966-1985

Two-year college enrollments total about 5,000,000. They decreased by 2% over the period 1980-1985.

During that five-year period, mathematical science course enrollments showed virtually the same percentage decrease. This is the first decrease we have observed in our regular surveys. See Graph 5-C for mathematical science enrollments.

GRAPH 5 - A

TRENDS IN OVERALL TWO-YEAR COLLEGE ENROLLMENTS, 1966-1985

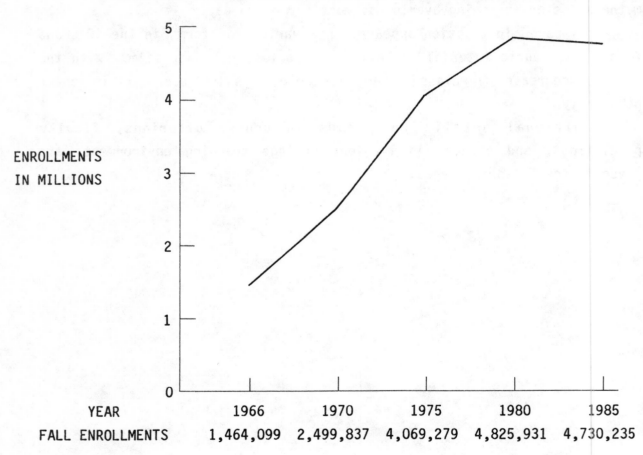

YEAR	1966	1970	1975	1980	1985
FALL ENROLLMENTS	1,464,099	2,499,837	4,069,279	4,825,931	4,730,235

Source: <u>1986 Community, Junior, and Technical College Directory</u>, AACJC, One Dupont Circle, N.W., Washington, D.C. 20036.

FULL-TIME VERSUS PART-TIME ENROLLMENTS IN TWO-YEAR COLLEGES, 1966-1985

Part-time enrollments overtook full-time enrollments in 1972. In 1985 part-time enrollments accounted for 65% of total enrollments.

GRAPH 5 - B

OVERALL FULL-TIME VERSUS PART-TIME ENROLLMENTS IN TWO-YEAR COLLEGES

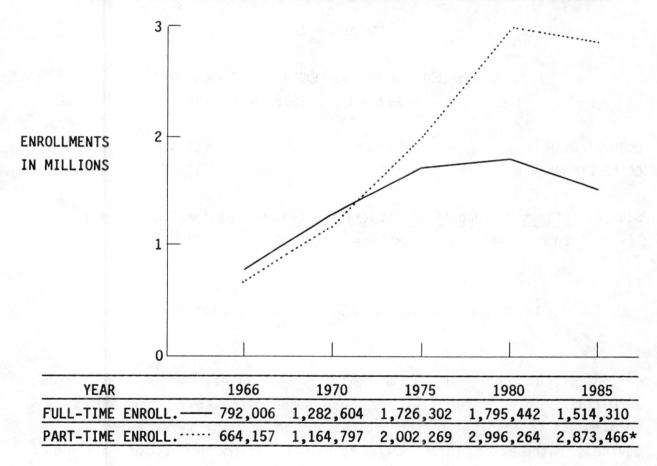

YEAR	1966	1970	1975	1980	1985
FULL-TIME ENROLL.———	792,006	1,282,604	1,726,302	1,795,442	1,514,310
PART-TIME ENROLL.······	664,157	1,164,797	2,002,269	2,996,264	2,873,466*

* The sum of full-time and part-time enrollments does not agree with total enrollments given on the previous page because the AACJC totals include "non-respondent" projections.

Source: Community, Junior, and Technical College Directories 1967, 1972, 1976, 1981, and 1986.

GROWTH OF ASSOCIATE DEGREES IN OCCUPATIONAL PROGRAMS
IN TWO-YEAR COLLEGES, 1970-1985

Since 1973-74 associate degrees in occupational programs have outnumbered associate degrees in college transfer programs. According to Cohen*, students in occupational programs tend to graduate at approximately the same rate as students in other programs. However, some students who transfer to four-year colleges do not complete associate degrees before transferring.

TABLE 5 - 1

ASSOCIATE DEGREES IN TWO-YEAR COLLEGE PROGRAMS

	1970-71	1975-76	1980-81	1981-82
OCCUPATIONAL	42.6%	55.2%	62.6%	63.5%
COLLEGE TRANSFER	57.4%	44.8%	37.4%	36.5%

Source: Digest of Educational Statistics 1983-84, National Center for Educational Statistics, Washington, D.C., p. 137.

TRENDS IN TWO-YEAR COLLEGE MATHEMATICS ENROLLMENTS

A slight decrease in mathematics enrollments marked the period 1980-1985. This is the first decrease noted since CBMS began monitoring enrollments in 1966. The decrease was fueled by large percentage drops in business mathematics (down 42%), technical mathematics (down 53%), and mathematics for liberal arts (down 42%). Enrollments in mathematics for liberal arts are now one-half of the 1966 level.

Remedial courses continued to gain (up 9%) and now account for 47% of

*Arthur M. Cohen and Florence B. Brawer, The American Community College, Jossey Bass, San Francisco, 1982.

total enrollments. Calculus enrollments increased by 13% and statistics was up 29%. Computing course enrollments slowed dramatically, growing by only 3% in the period 1980-1985. Computing course enrollments are nearly equal to calculus enrollments.

In 1980 we observed: "Courses of an applied nature showed the largest percentage increase in enrollments over the period 1975-1980, reflecting the greatly increased occupational/technical focus of two-year colleges." Five years later, enrollments in applied courses slowed, with technical mathematics and business mathematics decreasing. The continuing decline in business mathematics, first noted in 1980, is puzzling. Business mathematics enrollments also decreased in divisions outside mathematics.

REMEDIAL COURSES

Since 1966, the growth of remedial courses has been large indeed. In fact, the remedial course group (arithmetic, general mathematics, elementary algebra, intermediate algebra, and high school geometry) now accounts for nearly one-half of all tyc mathematics enrollments. This growth has alarmed many individuals who are concerned about tyc mathematics.

In spite of the large overall enrollments, there is an indication that some improvement is occurring at the pre-algebra level (arithmetic and general mathematics): Over the period 1980-1985, pre-algebra enrollments decreased by 3%, the first decrease noted since 1966.

ENROLLMENT TRENDS IN MATHEMATICAL SCIENCE COURSE GROUPS
1966-1985

Overall enrollments in mathematics courses decreased by 1% from 1980-1985 and thus mirrored the overall enrollment decrease of 2% in two-year colleges.

GRAPH 5 - C

MATHEMATICS COURSE ENROLLMENTS OVER TIME

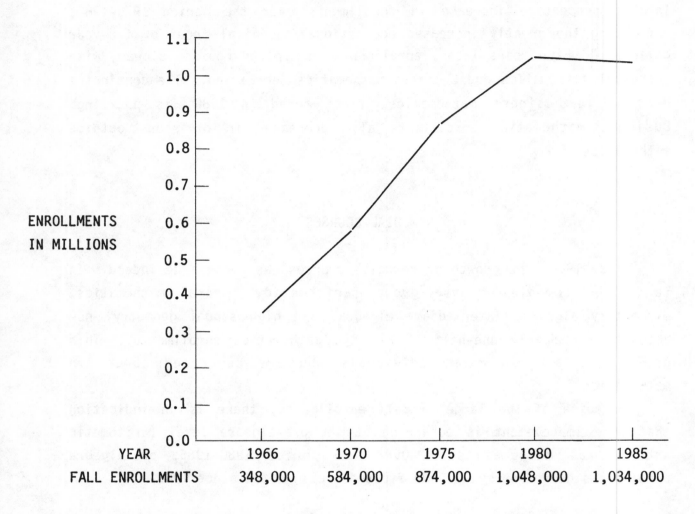

YEAR	1966	1970	1975	1980	1985
FALL ENROLLMENTS	348,000	584,000	874,000	1,048,000	1,034,000

(Y-axis: ENROLLMENTS IN MILLIONS, 0.0 to 1.1)

Table 5-2 gives enrollment trends by various courses and Graph 5-D, percentage trends in various course groups. Remedial course enrollments continued to grow over the 1980-1985 period, but their rate of growth decreased. Since 1980 the percentage shares of calculus, precalculus, and statistics have remained nearly level.

The computing boom of 1975-1980 seems to be over. Course enrollments in computing (including data processing enrollments) are only slightly higher than they were in 1980.

118

TABLE 5 - 2

DETAILED FALL ENROLLMENTS IN MATHEMATICAL SCIENCES IN TWO-YEAR COLLEGES, 1966-1985 (in thousands)

SUBJECT	1966	1970	1975	1980	1985
REMEDIAL					
1 Arithmetic	15	36	67	121	77
R General Mathematics	17	21	33	25	65
3 Elementary Algebra	35	65	132	161	181
4 Intermediate Algebra	37	60	105	122	151
5 High School Geometry	5	9	9	12	8
PRECALCULUS					
6 College Algebra	52	52	73	87	90
7 Trigonometry	18	25	30	33	33
8 College Alg. & Trig. (Combined)	15	36	30	41	46
9 Elementary Functions	7	11	16	14	13
CALCULUS					
10 Analytic Geometry	4	10	3	5	6
11 Analytic Geometry & Calculus	32	41	40	45	49
12 Calculus (math., physics & engr.)	8	17	22	28	31
13 Calculus (bio., soc., & mgt. sci.)	NA*	NA	8	9	13
14 Differential Equations	2	1	3	4	4
SERVICE COURSES					
15 Linear Algebra	1	1	2	1	3
16 Discrete Mathematics	NA	NA	NA	NA	L*
17 Finite Mathematics	3	12	12	19	21
18 Mathematics for Liberal Arts	22	57	72	19	11
19 Mathematics of Finance	4	5	9	4	1
20 Business Mathematics	17	28	70	57	33
21 Math. for Elem. School Teachers	16	25	12	8	9
22 Elementary Statistics	4	11	23	20	29
23 Probability & Statistics	1	5	4	8	7
24 Technical Mathematics	19	26	46	66	31
25 Technical Math. (calculus level)	1	3	7	14	4
26 Use of Hand Calculators	NA	NA	4	3	6
COMPUTING					
27 Data Processing (elem. or adv.)	NA	NA	NA	NA	36
28 Elem. Progr. (BASIC, COBOL, FORTRAN, Pascal)	3	10	6	58	37
29 Advanced Programming	NA	NA	NA	NA	5
30 Assembly Language Programming	NA	NA	NA	NA	4
31 Data Structures	NA	NA	NA	NA	2
32 Other Comp. Sci. Courses	2	3	4	37	14
33 Other Mathematics Courses	8	14	32	27	14
TOTALS	348	584	874	1048	1034

*(NA means "not available" and L means some but less than 500.)

GRAPH 5 - D

FALL ENROLLMENTS IN SELECTED TYPES OF MATHEMATICAL SCIENCE COURSES IN TWO-YEAR COLLEGES, BY LEVEL (As Percent of Total)

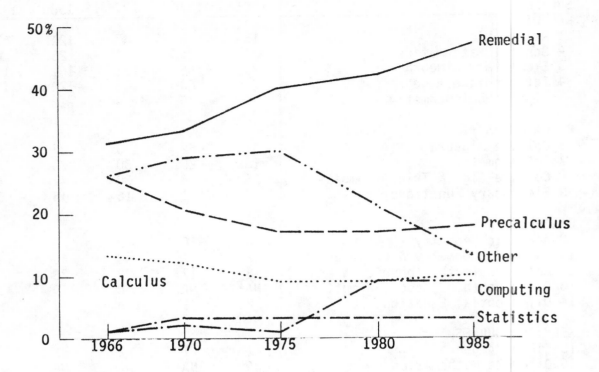

ENROLLMENTS IN THOUSANDS AND PERCENTAGES OF TOTAL

LEVEL	1966 NUMBER	%	1970 NUMBER	%	1975 NUMBER	%	1980 NUMBER	%	1985 NUMBER	%	
Remedial* (Courses 1-5)	109	31%	191	33%	346	40%	441	42%	482	47%	————
Precalculus** (6-9)	92	26%	124	21%	149	17%	175	17%	182	18%	— — —
Calculus (10-14)	46	13%	69	12%	76	9%	91	9%	103	10%	·········
Computing & D.P. (27-32)	5	1%	13	2%	10	1%	95	9%	98	9%	—·—··
Statistics (22-23)	5	1%	16	3%	27	3%	28	3%	36	3%	——·—
Other	91	26%	171	29%	266	30%	218	21%	133	13%	—··—
Total	348		584		874		1048		1034		

Note: This table was constructed using table on previous page. Percentages may not add to 100% due to rounding.

* Remedial courses include arithmetic, high school geometry, elementary algebra, intermediate algebra, and general mathematics (courses 1-5).

** Precalculus courses include college algebra, college algebra and trigonometry, trigonometry, and elementary functions (courses 6-9).

FIFTEEN YEAR TRENDS IN AVAILABILITY OF MATHEMATICS COURSES
1970-1985

Since 1970, remedial courses have become more widely available. In 1970, courses in arithmetic were taught in one-third of tyc's. In 1985, arithmetic was taught in more than one-half of tyc's. Calculus courses designed for engineering, science, mathematics, and physics are unchanged in availability since 1970. This steady availability may be explained in part by the introduction of new "soft" calculus courses designed for students in the biological, social, and managerial sciences. Soft calculus courses are available in 30% of tyc's.

Statistics is now taught in about three-fifths of tyc's; in 1970 it was taught in only two-fifths of tyc's.

The next table provides additional details on fifteen-year trends in availability. In contrast to the situation on availability of courses in four-year colleges where the questions asked were different in 1985 than in previous years (see the discussion preceding Table 1-7) the tyc questions for 1970 and 1985 seemed comparable. The results generally bear this judgment out. In the four-year college questionnaire, the issue was one of availability of upper division courses on a two-year cycle - an issue that hardly exists for two-year colleges where almost all courses would normally be taught every year.

TABLE 5 - 3

AVAILABILITY OF MATHEMATICAL SCIENCE COURSES IN TWO-YEAR COLLEGES
FIFTEEN-YEAR TRENDS, 1970-1985
Percentage of two-year colleges teaching the course

SUBJECT	FALL 1970	FALL 1985
REMEDIAL		
1 Arithmetic	37%	53%
2 General Mathematics	20%	41%
3 Elementary Algebra	48%	75%
4 Intermediate Algebra	56%	74%
5 High School Geometry	24%	18%
PRECALCULUS		
6 College Algebra	53%	76%
7 Trigonometry	64%	67%
8 College Alg. & Trig. (Combined)	41%	47%
9 Elementary Functions	25%	21%
CALCULUS		
10 Analytic Geometry	18%	17%
11 Analytic Geometry & Calculus	63%	58%
12 Calculus (math., physics & engr.)	41%	41%
13 Calculus (bio., soc., & mgt. sci.)	NA*	30%
14 Differential Equations	49%	40%
SERVICE COURSES		
15 Linear Algebra	17%	24%
16 Discrete Mathematics	NA	3%
17 Finite Mathematics	19%	27%
18 Mathematics for Liberal Arts	NA	25%
19 Mathematics of Finance	13%	5%
20 Business Mathematics	38%	34%
21 Math. for Elem. School Teachers	48%	31%
22 Elementary Statistics	41%	61%
23 Probability & Statistics	16%	15%
24 Technical Mathematics	41%	42%
25 Technical Math. (calculus level)	19%	18%
26 Use of Hand Calculators	NA	4%
COMPUTING		
27 Data Processing (elem. or adv.)	NA	28%
28 Elem. Progr. (BASIC, COBOL, FORTRAN, Pascal)	27%	46%
29 Advanced Programming	NA	19%
30 Assembly Language Programming	NA	12%
31 Data Structures	NA	5%
32 Other Comp. Sci. Courses	16%	16%

* (NA means not available - not gathered in 1970)

MATHEMATICS COURSES TAUGHT OUTSIDE OF MATHEMATICS PROGRAMS

We have previously noted the shift of two-year college enrollments to occupational programs. Many of these programs provide their own mathematics instruction. To get an approximation of the size of such "outside" offerings, we asked for estimates of enrollments in mathematics courses given by other divisions or departments. The estimates are probably not as reliable as other data presented in this report, because respondents did not have direct responsibility for these outside courses.

The majority of outside enrollments are found in computer science courses, data processing, and business mathematics. The divisions providing most of the outside courses are those which specialize in business and occupational programs.

In 1967, Jewett and Lindquist observed that "...The mathematics curriculum in junior colleges seems overwhelmingly designed for transfer students." Their words take on added importance in view of the continuing growth of occupational programs. Outside enrollments in mathematics and computer science, primarily in such programs, have nearly quadrupled since 1970 and are now estimated to be 35% of mathematics enrollments in mathematics programs. Without data processing, the estimate would be 20%.

Trends in "outside" enrollments had some parallels with "inside" enrollments: business mathematics and technical mathematics decreased and computing courses demonstrated little change from 1980. Other trends may be seen in Tables 5-4 and 5-5.

In 1985, computer science and data processing are the most prominent courses for "outside" mathematics enrollments. Computer science accounts for 27% of "outside" enrollments, decreasing slightly from 1980. "Outside" enrollments in business mathematics have decreased by 29% since 1980. "Inside" business mathematics enrollments also decreased, but by 42%. "Data processing" was not listed on previous surveys and may have been interpreted by some as "computer science and programming." If data processing is deleted, "outside" enrollments would have shown a decrease of 23%. However, some data processing may have been included in computer science totals prior to 1985.

TABLE 5 - 4

ESTIMATED ENROLLMENTS IN MATHEMATICS COURSES TAUGHT OUTSIDE
OF MATHEMATICS PROGRAMS IN TYC'S, FALL 1985
(Enrollments in Thousands)

COURSES	1970	1975	1980	1985
Arithmetic	14	27	18	18
Business Mathematics	36	53	70	50
Calculus or Differential Eqns.	L*	4	8	L
Computer Science & Programming	21	51	92	97
Data Processing	NA*	NA	NA	159
Pre-Calculus Coll. Mathematics	6	17	29	3
Statistics and Probability	6	14	12	7
Technical Mathematics	NA	NA	25	23
Other	9	12	10	4
Total	92	178	264	361

* L denotes some but less than 500 and NA denotes not available.

DIVISIONS OTHER THAN MATHEMATICS THAT TAUGHT
MATHEMATICS COURSES, FALL TERM 1985-1986

Business and occupational program faculties teach substantial numbers of mathematics courses.

TABLE 5 - 5

ENROLLMENTS IN COURSES IN OTHER DIVISIONS
(Enrollment in Thousands)

COURSES	NATURAL SCIENCES	OCCUPAT. PROGRAMS	BUSINESS	SOCIAL SCIENCES	OTHER	TOTAL
Arithmetic	L*	10	3	0	4	18**
Bus. Mathematics	0	4	46	0	L	50
Statistics & Prob.	0	0	4	2	L	7**
Pre-Calculus College Math.	0	3	0	0	0	3
Calculus or Diff. Eqns.	0	L	0	0	0	L
Comp. Sci. & Prog.	L	27	44	0	26	97
Data Processing	3	25	93	0	37	159**
Technical Math.	L	23	L	0	0	23
Other	0	4	0	0	0	4
Total	3	96	190	2	67	361**

* L denotes some but less than 500.

** denotes disagreement due to rounding.

COMPUTERS AND CALCULATORS IN TWO-YEAR COLLEGES

The percentage of two-year colleges reporting access to computers has increased from 57% in 1975 and now amounts to 84% of all tyc's. The mean number of computer terminals and microcomputers available for student use in mathematics courses is 19, with a median of 13. Department heads estimate that 59% of the full-time faculty know a computer language, the same percentage as in 1980. The number of faculty making use of computers in their teaching has doubled since 1975 and 32% of full-time faculty give some class assignments involving the use of the computer each year (in courses other than computer science). This figure is up from 21% in 1980. The impact of computers on mathematics teaching is growing but is still small; less than 7% of all sections of mathematics (excluding computer science) reported the use of computer assignments for students.

The impact of hand calculators on mathematics teaching is substantially larger than that of computers. Hand calculators are recommended for use in 43% of all course sections, up from 29% in 1980. Then, usage of calculators was concentrated in a small number of courses. Only courses in college algebra and trigonometry, trigonometry, statistics, and technical mathematics had usage rates in excess of 50%. (That is, the fraction of sections in which hand calculators was recommended exceeded 50%.) In 1985, 13 courses had usage rates over 50%: analytic geometry and calculus, business mathematics, calculus, college algebra and trigonometry, elementary functions, finite mathematics, mathematics for liberal arts, mathematics of finance, probability and statistics, soft calculus, statistics, technical mathematics (calculus level), and trigonometry.

INSTRUCTIONAL FORMATS FOR TWO-YEAR COLLEGE MATHEMATICS

The 1975 CBMS survey of two-year college mathematics noted the emergence of a variety of self-pacing instructional methods. The 1980 survey showed continued growth in use of self-pacing methods. The 1985 survey reveals a marked decrease in the use of self-pacing methods. The simplest explanation for this change is the decrease in teaching demands of faculty. From 1980-85 mathematical science enrollments decreased by 1% and the size of the faculty increased by 12%. Some would argue that the use of self-pacing methods increased, in part, during the 70's as a result of overloaded classrooms (and teachers).

For each of eleven instructional methods, the table below shows the percentage of two-year colleges reporting no use, use by some faculty, or use by most faculty of that instructional method in mathematics courses in 1985. The pronounced increase in the percentages of tyc's reporting no use of various alternative systems clearly shows the decline in popularity of all non-traditional instructional methods.

TABLE 5 - 6

INSTRUCTIONAL FORMATS

INSTRUCTIONAL METHOD	Not Being Used		Used By Some Faculty		Used By Most Faculty	
	1980	1985	1980	1985	1980	1985
Standard Lecture-Recit. Sys. (class size < or = to 40)	1%	1%	2%	14%	97%	85%
Large Lecture Classes (>40) with recitation sections	63%	77%	16%	19%	21%	4%
Large Lecture Classes (>40) with no recitation	76%	81%	12%	17%	12%	1%
Organized Program of Independent Study	37%	61%	62%	38%	1%	2%
Courses by Television (closed-circuit or broadcast)	73%	91%	27%	9%	0%	0%
Courses by Film	75%	96%	24%	4%	1%	0%
Courses by Programmed Instruc.	40%	69%	56%	27%	4%	4%
Courses by Computer-Assisted Instruction (CAI)	68%	74%	31%	24%	1%	2%
Modules	42%	68%	54%	25%	4%	6%
Audio-Tutorial	55%	75%	43%	24%	2%	2%
PSI (Personalized System of Instruction)	51%	76%	46%	20%	3%	4%

USE AND STAFFING OF MATHEMATICS LABORATORIES IN TWO-YEAR COLLEGES

Mathematics labs (math help centers, math tutorial centers) are relatively new adjuncts to mathematics instruction in two-year colleges. They may contain some or all of the following: tutors, calculators, computers, films, film strips, television units for playback of lectures or video cassettes, models, audio-tape units, learning modules, etc. Math labs have been established at a fairly constant rate since 1970 and can now be found in 82% of all two-year colleges, up from 68% in 1980. As shown in the table below, personnel of labs come from a variety of sources.

TABLE 5 - 7

SOURCES OF PERSONNEL FOR MATHEMATICS LABORATORIES

	Percent of TYC's Using Source
Students	48%
Full-time Members of Mathematics Staff	38%
Paraprofessionals	34%
Part-time Members of Mathematics Staff	30%
Members of Other Departments	19%
Other	3%

COORDINATION OF COLLEGE-TRANSFER PROGRAMS WITH FOUR-YEAR INSTITUTIONS

For two-year colleges with large degree-credit programs it is important to coordinate program offerings, advisement, and academic standards with the most likely four-year college or university destination of their students. Sixty-six percent of the responding tyc's reported that their mathematics offerings are subject to state regulation, and thirty-five percent reported official state-wide coordination of tyc mathematics offerings with those of four-year institutions.

This may help to explain the low level of reported consultation of tyc mathematics departments with four-year college and university departments: less than once a year for thirty-five percent, yearly for forty-one percent, and more than once a year for twenty-three percent.

CHAPTER 6

MATHEMATICAL SCIENCE FACULTY IN TWO-YEAR COLLEGES

This chapter describes the number, educational qualifications, professional activities, and selected personal characteristics of two-year college mathematical science faculty. For two-year colleges the terms "mathematical science" and "mathematics" describe the same faculty and are used interchangeably in that context. There is generally no separate computer science faculty. Computer science type courses are taught in many mathematics departments or divisions and as shown in Table 5-5, are also widely taught in occupational and business type programs. See the questionnaire, Appendix C, for interpretation of "mathematics department". The chapter includes profiles of the age, sex, and ethnic composition of mathematics faculty and information on mobility into, within, and out of two-year college teaching positions. Also included is a section on the teaching environment of mathematics faculty. While, prior to the 1980 report separate profiles were given for public and private tyc faculties, in the 1980 report the two faculties were combined, since only about 5% of the total faculty was in private tyc's. We continue this pattern started in 1980.

HIGHLIGHTS
1980-85

■ The full-time mathematics faculty increased by 12% since 1980 and now numbers 6,277.

■ The part-time mathematics faculty also increased by 12% and now numbers 7,433. Since 1980, part-timers have accounted for more than one-half of the total mathematics faculty.

■ The percentage of doctorates on the mathematics faculty decreased from 15% to 13% of the total, the first decrease noted by CBMS since 1970.

■ The percentage of mathematics faculty having highest degrees in computer science increased from 3% to 8%.

■ The percentage of mathematics faculty having highest degrees in statistics increased to 3%.

■ Women on the mathematics faculty increased to 31%, a gain of 10 percentage points in ten years.

■ Ethnic minorities on the mathematics faculty increased to 12%, up from 8% in 1975.

■ Overload teaching, usually for extra pay, remains prominent among tyc mathematics faculty, with 43% of faculty reported as teaching overloads.

■ Standard teaching loads decreased for the first time since 1970.

■ Remediation was cited as the biggest problem facing mathematics departments in the mid-80's.

NUMBER AND EDUCATIONAL QUALIFICATIONS OF TWO-YEAR COLLEGE FACULTY

As of fall 1985, two-year colleges employed 93,611 full-time faculty and 135,195 part-time faculty. More than 75% hold a master's degree and 14% hold a doctorate. Since two-year colleges emphasize teaching and not research, two-year college faculty spend significantly more time in the classroom than do faculty in four-year colleges and universities. Most two-year college faculty teach about 16 hours per week.

Since more than 50% of all students enrolled at two-year colleges are taking courses in occupational fields, faculty trained and experienced in such areas as health technologies, business, data processing, and public service fields and disciplines that serve these fields are currently in greatest demand. Even so, our survey results show that the growth, since 1980, of the full-time equivalent (FTE) mathematics faculty was 12%, in marked contrast to the 7% decrease of all two-year college faculty shown in Graph 6-A.

GRAPH 6 - A

NUMBERS OF FULL-TIME EQUIVALENT (FTE) TYC FACULTY, ALL FIELDS
(In Thousands)

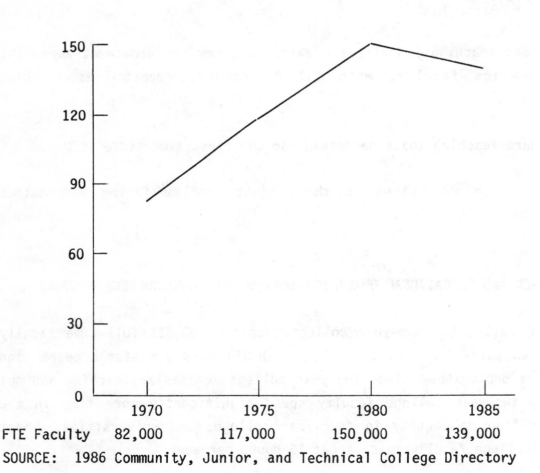

	1970	1975	1980	1985
FTE Faculty	82,000	117,000	150,000	139,000

SOURCE: 1986 Community, Junior, and Technical College Directory

TRENDS IN NUMBERS OF FULL- AND PART-TIME MATHEMATICS FACULTY

For mathematics in two-year colleges, part-time faculty now outnumber full-time faculty, making up 54% of the total. For _all_ fields in tyc's, part-timers constitute 59% of the faculty. The part-time component of the mathematics faculty increased by 12% over the period 1980-1985, down sharply from the 95% increase observed in 1975-1980. The 12% increase in the size of the full-time faculty matched the increase of the part-time faculty.

GRAPH 6 - B

FULL- AND PART-TIME MATHEMATICS FACULTY DISTRIBUTION OVER TIME
(In Thousands)

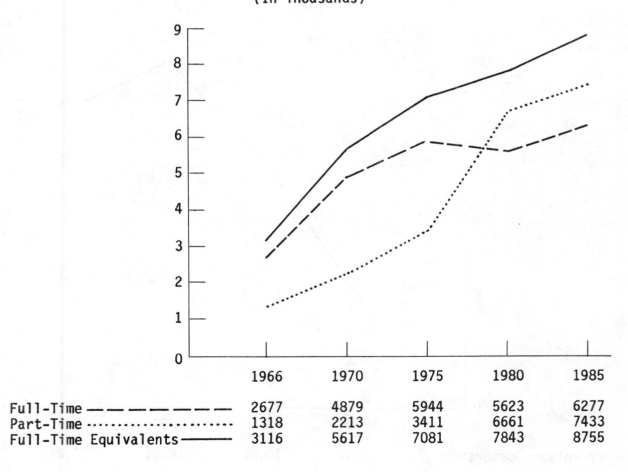

	1966	1970	1975	1980	1985
Full-Time	2677	4879	5944	5623	6277
Part-Time	1318	2213	3411	6661	7433
Full-Time Equivalents	3116	5617	7081	7843	8755

TRENDS IN DOCTORATES AMONG FULL-TIME MATHEMATICS FACULTY

The percentage of doctorates among the full-time mathematics faculty in two-year colleges declined over the period 1980-1985. This ends a period of steady growth in the percentage of doctorates on mathematics faculty. The current figure of 13% is close to the 14% figure of doctorates on the total tyc faculty. The percentage of doctorates on the four-year college and university mathematical and computer science faculty also decreased over the period 1980-1985.

GRAPH 6 - C

PERCENTAGE OF DOCTORATES AMONG FULL-TIME MATHEMATICS FACULTY

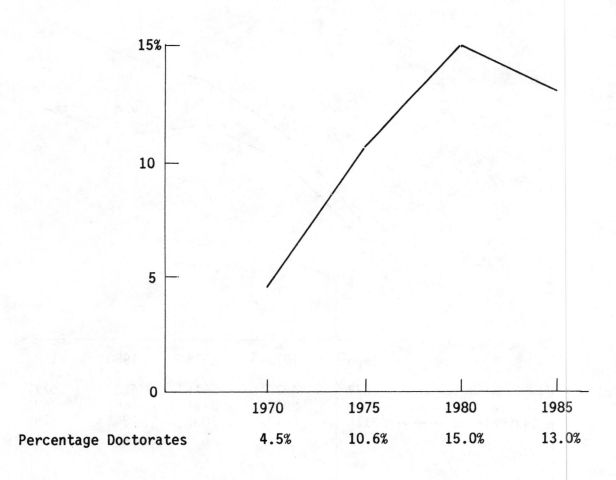

	1970	1975	1980	1985
Percentage Doctorates	4.5%	10.6%	15.0%	13.0%

HIGHEST ACADEMIC DEGREES OF FULL-TIME MATHEMATICS FACULTY 1985

Table 6-1 gives the percentages of the total tyc mathematics faculty by field of highest degree and the level of that training. Since 1980 the percentages of the faculty holding highest degrees in statistics and in computer science went up markedly, from 1% to 3% in statistics and from 3% to 8% in computer science. But note that fewer than one-half of one percent had doctorates in these areas. Except for increases at the non-doctorate levels in computer science and statistics, the overall matrix for 1985 is very similar to that for 1980. The degree level "Masters +1" (or Mast. +1) refers to one year beyond the Masters level.

TABLE 6 - 1

TYC FULL-TIME MATHEMATICS FACULTY BY FIELD AND LEVEL OF HIGHEST DEGREE

Degree Level

Field	Doct.	Mast.+ 1	Masters	Bachelors	Total
Mathematics	6%	26%	24%	2%	58%
Statistics	0%	1%	1%	1%	3%
Computer Science	0%	3%	4%	1%	8%
Mathematics Educ.	4%	6%	8%	0%	18%
Other Fields	3%	3%	6%	1%	13%
Total	13%	39%	43%	5%	100%

135

TABLE 6 - 2

DEGREE STATUS OF FULL-TIME TYC MATHEMATICS FACULTY, 1970-1985
(As Percent of Total Full-Time Mathematics Faculty)

Highest Degree	1970	1975	1980	1985
Doctorate	4%	11%	15%	13%
Masters + 1 year	47%	35%	38%	39%
Masters	42%	47%	42%	43%
Bachelors	7%	7%	5%	5%

AGE, SEX, AND ETHNIC COMPOSITION OF TWO-YEAR COLLEGE
MATHEMATICS FACULTY

Since 1980 the full-time faculty in mathematics has increased by 12% at a time when there has been a percentage decrease in the group under age 40 and a percentage increase in the 40-49 age group. There are continuing indications that a substantial number of faculty in the over 45 age group are leaving two-year college mathematics teaching.

During the ten-year period 1975-85, the female percentage of two-year college full-time mathematics faculty has risen from 21% to 31%, with a numerical increase in the number of female faculty from 1,250 in 1975 to 1,946 in 1985. From 1980 to 1985, the rate of growth of the number of women on the two-year college mathematics faculty was three times the growth rate of the overall mathematics faculty.

Ethnic minorities have continued to increase, from 8% of the total faculty in 1975 to 12% in 1985.

TRENDS IN AGE DISTRIBUTION OF FULL-TIME MATHEMATICS FACULTY, 1975-1985

Trite as it may sound, the full-time tyc mathematics faculty is not getting any younger. In 1975, 47% of the faculty was under 40 years of age; today the figure is 34%. Over the same ten year period, the percentage between 40 and 49 has increased from 28% to 42%. The percentage of faculty over 50 years of age has remained fairly steady.

In Table 6-3 the trends since 1975 of the age composition of the full-time faculty are shown. The diagonal arrows indicate the translation of an age group to the corresponding five year older group five years later. Clearly, hiring occurs up to age 45 or 50. The table also indicates early retirements or dropouts among faculty who were over 45 years of age in 1980.

TABLE 6 - 3

AGE DISTRIBUTION OF FULL-TIME TYC MATHEMATICS FACULTY

Age Range	Percent of Full-Time Mathematics Faculty			Number of Full-Time Mathematics Faculty			Change: 1980-1985
	1975	1980	1985	1975	1980	1985	
< 30	9%	5%	5%	535	281	314	314
30-34	18%	15%	11%	1070	843	690	409
35-39	20%	24%	18%	1188	1350	1130	287
40-44	15%	18%	24%	892	1012	1506	156
45-49	13%	16%	18%	773	900	1130	118
50-54	13%	10%	13%	773	562	816	-84
55-59	8%	7%	7%	475	394	439	-123
60 or more	4%	5%	4%	238	281	252	-142
Totals				5944	5623	6277	

AGE DISTRIBUTION OF FULL-TIME MATHEMATICS FACULTY BY SEX, 1985

From 1975 to 1985 the number of women on full-time mathematics faculties of two-year colleges has increased from 21% to 31% of the total. As might be expected, women are more heavily represented in younger age groups, with more than one-fourth less than 35 years of age. Only 28% of female faculty members are 45 or more years of age as contrasted to 48% of male faculty members. The total number of men is 4,331 and the total number of women is 1,946.

TABLE 6 - 4

1985 AGE DISTRIBUTION OF FULL-TIME FACULTY BY SEX

Age Range	Male	Female
< 35	13%	26%
35-44	40%	45%
45-54	36%	19%
55 or more	12%	9%

Totals may not be 100% due to rounding.

ETHNIC GROUPS AMONG FULL-TIME MATHEMATICS FACULTY, 1985

The ethnic-group distribution of the full-time mathematics faculty of two-year colleges in 1985 is shown in the Table 6-5. The total minority-group fraction is now 12%, up from 8% in 1975. Hispanics registered the greatest gains. (The total number of non-Caucasian ethnic group faculty is 753.)

TABLE 6 - 5

1985 ETHNIC GROUP DISTRIBUTION OF FULL-TIME FACULTY

Ethnic Group	Percentage of Total
Caucasian	88%
Asian	3%
Hispanic	4%
Black	4%
American Indian	1%

The age distribution of the ethnic minority part of the full-time mathematics faculty of two-year colleges in 1985 is shown in Table 6-6. It differs from the overall faculty age distribution shown in Table 6-3 primarily in having a larger fraction under age 35 and a smaller fraction of age 55 or over, but is quite similar to the female faculty age distribution shown in Table 6-4.

TABLE 6 - 6

1985 AGE DISTRIBUTION OF ETHNIC MINORITY FACULTY

Age Range	Percent of Total Ethnic Minority Faculty
< 30	27%
35-44	46%
45-54	20%
55 or more	7%

PART-TIME MATHEMATICAL SCIENCE FACULTY IN TWO-YEAR COLLEGES

The part-time faculty now numbers 7,433 and increased by 12% over the period 1980-1985, down sharply from a 95% increase in 1975-1980. Overall, for all fields, part-timers account for 59% of the two-year college faculty. Mathematics, until the year 1980, used part-timers more sparingly than did other departments, but now the part-time fraction is 54%. For all intents and purposes, mathematics faculty now have the dubious distinction of being on a vertical par with other departments.

As compared with the 1970 figures, the percentages of part-time mathematics faculty in the doctorate or "masters + 1" highest degree categories have declined. During the same fifteen-year period, the percentage of part-timers in the bachelors category has doubled and is now more than one-fourth of the total. Given an increase in the number of industrial opportunities for mathematicians, it is not likely that the educational qualifications of part-timers will soon increase.

TABLE 6 - 7

DEGREE STATUS OF PART-TIME MATHEMATICS FACULTY SINCE 1970
(As Percentage of Total Part-time Mathematics Faculty)

Highest Degree	1970	1975	1980	1985
Doctorate	9%	4%	7%	7%
Masters + 1	31%	30%	18%	15%
Masters	46%	49%	58%	50%
Bachelors	14%	17%	17%	28%

HIGHEST ACADEMIC DEGREES OF PART-TIME MATHEMATICS FACULTY, 1985

As might be expected, the degree qualifications of the full-time faculty exceed those of the part-time faculty. Compare Table 6-8 below with Table 6-1.

TABLE 6 - 8

TYC PART-TIME MATHEMATICS FACULTY BY FIELD AND LEVEL OF HIGHEST DEGREE

	Percent with Highest Degree			
Field	Doctorate	Masters + 1	Masters	Bachelors
Mathematics	3%	8%	30%	17%
Statistics	1%	0%	0%	0%
Computer Science	0%	1%	1%	3%
Mathematics Educ.	1%	4%	7%	3%
Other Fields	2%	2%	12%	5%
Totals	7%	15%	50%	28%

For 1985, high school teachers constitute the largest source of part-time mathematics faculty in two-year colleges, as shown in Graph 6-D.

GRAPH 6 - D

SOURCES OF PART-TIME FACULTY AS PERCENTAGES OF TOTAL

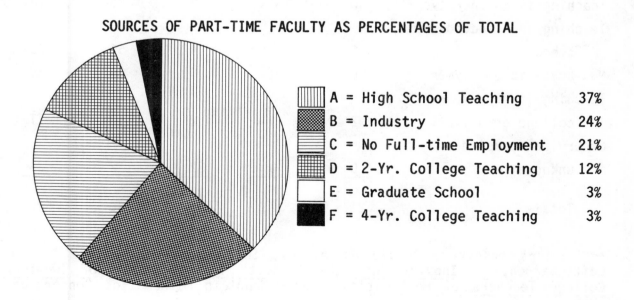

A = High School Teaching	37%
B = Industry	24%
C = No Full-time Employment	21%
D = 2-Yr. College Teaching	12%
E = Graduate School	3%
F = 4-Yr. College Teaching	3%

SOURCES OF NEW FULL-TIME MATHEMATICS FACULTY
IN TWO-YEAR COLLEGES, 1985

Twenty-nine percent of new full-time mathematics faculty in 1985 entered two-year college mathematics teaching directly from graduate school. Teaching part-time in a two-year college continues to be a viable path to full-time status, with 25% of new hires coming from that source. High schools seem to be a smaller source of new faculty than they were earlier. A 1979 survey showed that more than 60% of all mathematics faculty in two-year colleges had previously taught in secondary schools.*

TABLE 6 - 9

INFLOW OF NEW FULL-TIME MATHEMATICS FACULTY 1985

Source	- - - - Type of Doctorate - - - -				Totals
	Math.	Math. Ed.	Other	None	
Graduate School	17	0	2	134	153
Employed by same tyc	2	0	4	123	129
Teaching in another tyc	0	0	2	76	78
Teaching in a secondary school	0	7	4	59	70
Non-academic employment	0	0	0	39	39
Teaching in four-year college or univer.	2	0	2	13	17
Otherwise occupied or unknown	0	0	32	2	34
Totals	21	7	46	446	520

* Robert McKelvey, Donald J. Albers, Shlomo Liebeskind, and Don O. Loftsgaarden, An Inquiry into the Graduate Training Needs of Two-Year College Teachers of Mathematics, Rocky Mountain Mathematics Consortium, 1979.

FULL-TIME MATHEMATICS FACULTY LEAVING TWO-YEAR COLLEGES, 1985

The "death or retirement" category is at variance with the 1980 age distribution constructed by CBMS. The 1980 age distribution showed 5% of the faculty to be over 60 years of age. Assuming retirement at an average age of 65 that translates to approximately 56 retirements per year. Our total of 217 is about four times that estimate and suggests other phenomena at work, perhaps early retirements. A substantial portion of the 55-59 age group left two-year college teaching between 1980 and 1985. Many of them may be in the retiree group. In contrast to retirement conditions in four-year colleges and universities, many two-year colleges may have retirement systems like the those in public school systems, thereby presumably encouraging early retirements.

TABLE 6 - 10

OUTFLOW OF FULL-TIME MATHEMATICS FACULTY 1985

Source	Math.	Math. Ed.	Other	None	Totals
Died or retired	10	10	2	195	217
Teaching in 4-Yr. College or Univ.	10	0	5	47	62
Teaching in a sec. school	0	0	0	42	42
Non-academic employment	0	0	5	29	34
Teaching in a 2-Yr. College	10	0	0	18	28
Otherwise occupied or Unknown	0	0	0	66	66
Returned to Grad. school	0	0	0	0	0
Totals	30	10	12	397	449

Header spans: - - - - Type of Doctorate - - - - over Math., Math. Ed., Other, None

THE TEACHING ENVIRONMENTS OF MATHEMATICS FACULTY
IN TWO-YEAR COLLEGES

There is evidence in our CBMS Survey data that the teaching environments of two-year college mathematics faculty have improved since 1980. The bulk of that evidence is contained in the next three tables dealing with the number of students taught by an average faculty member, professional activities, and problems of the mid-80's. These tables tell us the following about two-year college mathematics faculty over the period 1980-85:

1. Student loads per FTE faculty have decreased.

2. Professional activity of faculty has increased.

3. There is a greatly increased concern about the use of part-time faculty for instruction and a heightened interest in maintaining vitality of faculty.

TRENDS IN STUDENT LOADS FOR TWO-YEAR COLLEGE MATHEMATICS FACULTY

Student loads have decreased sharply in two-year college mathematics programs, down by 16 students per FTE faculty member. In 1985, mathematics program heads reported that 43% of the full-time faculty were teaching overloads, usually one additional course (3 semester hours) beyond the standard load of 16 contact hours. Department heads reported that not all faculty teaching overloads received additional pay for such work. They, in fact, reported that 29% of faculty teaching overloads did not receive extra compensation. This overload faculty work might mask an undercount of the part-time share in FTE faculty time and thus overestimate the number of students per FTE faculty member. For the faculty actually teaching the overloads, the added responsibility means they must provide mathematics instruction for even more students.

144

The total number of sections taught by part-time faculty was 11,900, 28% of the total number of sections. This figure is supported by the FTE number of part-time faculty, 2,478, which is 28% of the total FTE faculty. (The average teaching load of full-time faculty is 16.1 contact hours and for part-timers it is 5.7). The ratio 5.7/16.1=0.35 provides support for the 1/3 conversion factor used in computing FTE (full-time equivalent faculty) numbers.

GRAPH 6 - E

MATHEMATICS ENROLLMENTS PER FTE FACULTY MEMBER

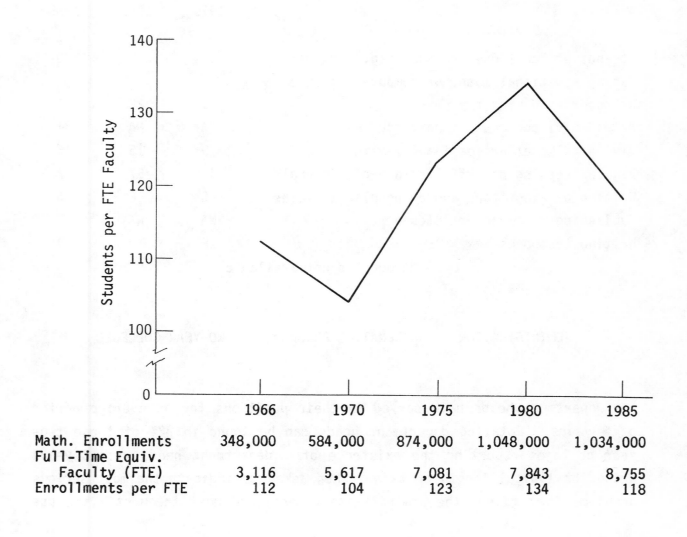

	1966	1970	1975	1980	1985
Math. Enrollments	348,000	584,000	874,000	1,048,000	1,034,000
Full-Time Equiv. Faculty (FTE)	3,116	5,617	7,081	7,843	8,755
Enrollments per FTE	112	104	123	134	118

PROFESSIONAL ACTIVITIES OF FULL-TIME MATHEMATICS FACULTY

Mathematics program heads in two-year colleges reported a continuing increase in professional activities of the faculty from 1975 to 1985. There is now more participation in conference attendance, reading of journals, and continuing education. Only textbook writing appears to have declined.

TABLE 6 - 11

PERCENT OF FACULTY ENGAGING IN ACTIVITY

Activity	1975	1980	1985
Attends at least one profess. mtg. per year	47	59	70
Taking additional math. or computer science courses during the year	21	22	31
Attend mini-courses or short courses	NA*	NA	31
Giving talks at professional meetings	9	15	16
Regular reading of articles in prof. journals	47	57	72
Writing of expository and/or popular articles	5	6	6
Publishing research articles	NA	NA	3
Writing textbooks	15	10	4

*NA denotes not available

ADMINISTRATION OF MATHEMATICS PROGRAMS IN TWO-YEAR COLLEGES

Department heads have served in their positions for an average period of 8 years. Rotating department heads can be found in 22% of those two-year colleges reporting the existence of a department head, with 3 years being the typical length of term. When asked to indicate the most serious problems they faced, the administrators mentioned most frequently the use of temporary faculty, "dealing with remediation", the use of part-time faculty, salary patterns, and problems related to computer facilities.

TABLE 6 - 12

PROBLEMS OF THE MID-80's

	Rank	Percent Classifying Problem As Major
The need to use temporary faculty for instruction	1	61%
Remediation	2	60%
Salary levels/patterns	3	53%
Computer facilities for classroom use	4	50%
Departmental support sources (travel funds, staff, secretary, etc.)	5	41%
Maintaining vitality of faculty	6	39%
Staffing computer science courses	7	34%
Upgrading/maintaining computer facilities	8	30%
Computer facilities for faculty use	9	27%
Class size	9	27%
Advancing age of tenured faculty	11	25%
Coordinating math. courses for four-year colleges and universities	12	22%
Classroom/lab facilities	13	21%
Coordinating and/or developing math. with voc./tech. programs	14	20%
Coordinating math. courses with sec. schools	15	19%
Office/lab facilities	15	19%
Library: holdings, access, etc.	17	7%
Lack of experienced senior faculty	17	7%
Losing full-time faculty to industry/government	17	7%

* Department heads used a six-point scale in rating the problems.
"Major problem" corresponds to an answer of 5 or 6 on the six-point scale.

APPENDIX A

SAMPLING AND ESTIMATION PROCEDURES

Sampling Procedure

The sampling frame was extracted from the 1982 National Center for Educational Statistics' HEGIS list which also gave Fall 1982 enrollments. The population frame consisted of those 2-year colleges, 4-year colleges or universities in the U.S.A., the District of Columbia or Puerto Rico that offered undergraduate Mathematics courses. There was a total of 2463 such institutions.

The population was divided into 22 strata on the basis of Control (Public or Private), Type (University, 4-year college or 2-year college) and Fall 1982 enrollment. This stratification is similar to but simpler than the one used for the 1980-81 CBMS study. Standard sampling techniques were used to determine the sample size for each stratum and then a random sample of institutions was chosen from each stratum.

Since the Stratification was based on enrollment, large schools were sampled much more heavily than small schools. Table A-1 gives a summary of the stratification.

Addresses were determined for sampled schools with the main sources of addresses being the 1985 Mathematical Sciences Administrative Directory and the 1985 Community, Technical, and Junior College Directory.

TABLE A - 1

NUMBER OF INSTITUTIONS IN EACH CONTROL/TYPE STRATUM
AND SAMPLE SIZE IN EACH STRATUM

Control/Type	# of Strata	Population	Sample
Public Universities	4	95	46
Private Universities	3	62	26
Public 4-Year Schools	4	427	105
Private 4-Year Schools	4	839	80
2-Year Schools	7	1040	172
Total	22	2463	429

   Appropriate questionnaires were sent to all Mathematics Departments in the sampled institutions or to the Division in charge of Mathematics courses. In addition questionnaires were mailed to all Computer Science, Statistics or other Mathematical Sciences Departments that were determined to exist at the sampled schools. Two-year colleges had a different questionnaire than the other schools. In addition, two short questionnaires dealing with remedial Mathematics and Computer Science were mailed to appropriate departments. The questionnaires are discussed in more detail in the report and copies of all questionnaires are found elsewhere in Appendices B to D.

   Table A-2 summarizes the population and sample sizes for the separate Computer Science and Statistics Departments in four-year colleges and universities.

TABLE A - 2

NUMBER OF COMPUTER SCIENCE AND STATISTICS DEPARTMENTS
IN POPULATION AND SAMPLE

Control/Type	Population*	Sample
Computer Science		
Universities	105	51
Public Colleges	141	40
Private Colleges	150	16
	396	107
Statistics		
Universities	40	21
Public Colleges	5	2
	45	23

*Population sizes are estimated from the sample.

Less than 10 Mathematical Sciences Departments other than Mathematics, Statistics and Computer Science were found in the sampled schools.

All projected enrollments in Mathematics courses and other information in this report are based on the information supplied by the Mathematical Sciences Departments mentioned earlier in this section. For example, no attempt was made to determine enrollments in Mathematics, Statistics or Computer Science courses that were taught in non-Mathematical or Computer Sciences Departments in four-year colleges or universities.

Estimation Procedures

Course enrollments and other information in this report are estimated national figures for all institutions in the frame described earlier in this Appendix for Fall 1985. The projections were made using standard procedures for stratified samples. For example, if stratum i has N_i schools in it, n_i schools respond with enrollments for course A and E_i is the total enrollment in Course A reported by these n_i schools, then the estimated total enrollment in Course A in Stratum i is given by:

$$\frac{N_i}{n_i} \cdot E_i$$

Required totals are then computed by adding estimates for appropriate strata.

The procedure used to handle separate Mathematical Sciences Departments at the same institution varied with the question. For course enrollments, data from all departments at each school were combined before projections were made. On the other hand, most information on faculty members was kept separate for the departments at each school.

Accuracy of Estimates

The response rates are given in Table A-3. They are down slightly from the 1980-81 study which had the highest response rates of any in this series of studies dating back to 1965-66.

TABLE A - 3

RESPONSE RATES IN DEPARTMENTS OF MATHEMATICS, STATISTICS, AND COMPUTER SCIENCE

	Pop.*	Sample	Respondents	Response Rate
1. Public Universities				
Mathematics	95	45	38	83%
Statistics	34	19	17	89%
Computer Science	78	39	24	62%
2. Private Universities				
Mathematics	62	26	18	69%
Statistics	6	2	2	100%
Computer Science	27	12	8	67%
3. Public Four-Year Colleges				
Mathematics	427	105	81	77%
Statistics	5	2	2	100%
Computer Science	141	40	24	60%
4. Private Four-Year Colleges				
Mathematics	839	80	57	71%
Computer Science	150	16	8	50%
5. Two-Year Colleges	1040	172	110	64%

SUMMARY BY DEPARTMENT

	Pop.*	Sample	Respondents	Response Rate
Mathematics	1423	257	194	75%
Statistics	45	23	21	91%
Computer Science	396	107	64	60%
	1864	387	279	72%

*Figures for Statistics and Computer Science Departments were estimated from the sample.

Followup phone calls were made to all departments not responding by a certain date as was done in earlier studies. Later when the statistical analysis was carried out, selected projections were made using only the first 60% of the questionnaires to be returned. These results agreed very well with the results for the entire data set.

The population frame (discussed earlier) had Fall 1982 enrollments for all schools. These enrollment figures for the responding schools were used to project total enrollments for all schools in the population. Actual enrollments were found by adding enrollments for all schools. Table A-4 contains a comparison of these results.

TABLE A - 4

COMPARISON OF ACTUAL AND ESTIMATED ENROLLMENTS

	Estimated Enrollment	Actual Enrollment	Error
Universities	2,866,665	2,903,490	-1.27%
Public Four-Year Colleges	3,026,499	2,978,696	+1.60%
Private Four-Year Colleges	1,515,073	1,582,379	-4.25%
Two-Year Colleges	4,810,920	4,642,187	+3.63%

A list of all responding departments is included as Appendix F.

APPENDIX B

FOUR-YEAR COLLEGE AND UNIVERSITY QUESTIONNAIRE
(SEE PAGE B-8 FOR REMEDIAL QUESTIONNAIRE)

SURVEY OF UNDERGRADUATE PROGRAMS

IN

THE MATHEMATICAL AND COMPUTER SCIENCES

1985

General Instructions

You are asked to report on programs in the mathematical and computer sciences (including statistics) under the cognizance of your department. This same questionnaire is being sent to each department in the mathematical or computer sciences on your campus which is listed in the 1985 Mathematical Sciences Professional Directory published by the AMS. It is not being routinely sent to computer centers or to non-departmental groups or programs listed there. Do not include data for branches or campuses of your institution that are geographically or budgetarily separate. Questions 1-9 are generally quantitative and non-judgmental in nature. Questions 10-13 involve more qualitative answers.

Please return completed questionnaire by 27 November 1985 to:

Conference Board of the Mathematical Sciences
1529 Eighteenth Street, N.W.
Washington, D.C. 20036
(202) 387-5200

* * *

For Coding Only

1. Name of your institution: _____ _ _ _ 16-18

Name of your department: _____ _ 19

2. Changes in Administrative Structure:

    (a) Between 1980 and 1985 was your department, together with one or more other departments, consolidated into a larger administrative unit (e.g., a Division of Mathematical Sciences or Department of Electrical Engineering and Computer Science)?   Yes ___ No ___   _ 20

    Names of other departments involved in this consolidation _____

    _____

    Name of larger administrative unit _____

    (b) Between 1980 and 1985 was your department divided with part of your faculty entering a new department (e.g., a new department of Statistics or Computer Science?)   Yes ___ No ___   _ 21

    Name of new department(s) _____

    (c) If you answered no to (a) and (b), was your present department created since 1980?   Yes ___ No ___   _ 22

    (d) Other major changes in administrative structure. Please specify:

    _____

    _____

3. Regular Undergraduate Program Courses, Fall 1985

Instructions for Question 3:

(a) The undergraduate courses in column (1) in the following table are listed in three groups corresponding roughly to a division into mathematics, statistics, and computer science. Within each group they are listed in approximate "catalog order" for your convenience in locating a listing which is a reasonable approximation to your offerings. Additional blank spaces are provided to permit you to write in names of courses which do not fit reasonably under some listed title.

For the purpose of this survey, consider as a single course, instruction in a particular area of mathematics which you offer as a sequence of two or more parts (e.g., calculus). Column (3) is to be used to indicate the number of sections of a course.

(b) For each course in column (1) that is being taught in the Fall of 1985 write in column (2) the total number of students who are enrolled in (any part of) the course this Fall. Thus, for a 2-semester sequence of calculus, Math 1 and Math 2, the enrollment in column 2 would be the sum of the Fall enrollments in Math 1 and in Math 2. Enter in column (3) the total number of sections of the course in the Fall of 1985. If a course is not being taught in the Fall but is expected to be taught during some other session of the current academic year, or was taught during the 1984-85 academic year, write S (for "sometimes") in column (2). If not taught during either of the academic years 1984-85 or 1985-86, write N. Each box in column (2) should contain an entry.

Undergraduate Courses

Name of Course (or equivalent) (1)	Total Number of Students Enrolled Fall 1985 (2)	Total Number of Sections (3)	
A. MATHEMATICS			
Remedial			
1. Arithmetic			23-28
2. General Math. (basic skills, operations)			29-34
3. Elem. Algebra (High School)			35-40
4. Intermed. Algebra (High School)			41-46
Pre-calculus			
5. College Algebra			47-52
6. Trigonometry			53-58
7. Coll. Algebra & Trig., combined			59-64

Name of Course (or equivalent)	Students Enrolled	Number of Sections	
8. Elem. Functions, Pre-calc. math.			65-70
9. Math. for Liberal Arts			71-76
10. Finite Mathematics			6-11
11. Business Mathematics			12-17
12. Math. for Elem. School Teachers			18-23
13. Analytic Geometry			24-29
14. Other Pre-calc.			30-35
Calculus Level			
15. Calc. (math., phys. sci., & engin.)			36-41
16. Calc. (bio., social & mgmt. sciences)			42-47
17. Differential Equations			48-53
18. Discrete Mathematics			54-59
19. Linear Alg. and/ or Matrix Theory			60-65
Advanced Level			
20. Modern Algebra			66-71
21. Theory of Numbers			72-77
22. Combinatorics			6-11
23. Graph Theory			12-17
24. Coding Theory			18-23
25. Foundations of Math.			24-29
26. Set Theory			30-35
27. Discrete Structures			36-41
28. History of Mathematics			42-47
29. Geometry			48-53
30. Math. for Sec. Sch. Teachers (methods, etc.)			54-59

Mathematics (continued)

Name of Course (or equivalent)	Students Enrolled	Number of Sections	
31. Mathematical Logic			60-65
32. Advanced Calculus			66-71
33. Advanced Math. for Eng. & Phys.			72-77
34. Vector Analysis, Linear Algebra			6-11
35. Advanced Diff. Equations			12-17
36. Partial Diff. Equations			18-23
37. Numerical Analysis			24-29
38. Applied Mathematics, Math. Modelling			30-35
39. Operations Research			36-41
40. Complex Variables			42-47
41. Real Analysis			48-53
42. Topology			54-59
43. Senior Seminar/Independ. Stud. Math.			60-65
44. Other Mathematics			66-71

3.A. TOTAL NO. OF MATHEMATICS SECTIONS — 72-74

B. STATISTICS

Name of Course (or equivalent)	Students Enrolled	Number of Sections	
45. Elem Stat. (no Calc. prereq.)			6-11
46. Prob'y (& Stat.)(no Calc. prereq.)			12-17
47. Math. Stat. (Calculus)			18-23
48. Probability (Calculus)			24-29
49. Stochastic Processes			30-35
50. Applied Stat. Analysis			36-41
51. Design & Analysis of Experiments			42-47
52. Regression (and Correlation)			48-53
53. Senior Seminar/Independ. Stud. Stat.			54-59
54. Other Statistics			60-65

3.B. TOTAL NO. OF STATISTICS SECTIONS — 66-68

C. COMPUTER SCIENCE

Name of Course (or equivalent) (1)	Total Number of Students Enrolled Fall 1985 (2)	Total Number of Sections (3)	
Lower Level			
55. Computers & Society			6-11
56. CS1 '78 or CS1 '84 (Comp. Prog. I) [*,**]			12-17
57. CS2, '78 (Comp. Prog. II) [*]			18-23
58. CS2, '84 [**]			24-29
59. Database Mgmt. Systems			30-35
60. Discrete Mathematics			36-41
61. Other lower level service courses			42-47
Middle Level			
62. Intro. to Comp. Systems (CS3)			48-53
63. Assembly Lang. Programming			54-59
64. Intro. to Comp. Organization (CS4)			60-65
65. Intro. to File Processing (CS5)			66-71
Upper Level			
66. Operating Sys. & Comp. Architect.			72-77
67. Operating Systems			6-11
68. Computer Architecture			12-17
69. Data Structures (CS7)			18-23
70. Survey of Prog. Languages			24-29
71. Computers & Society (CS9)			30-35
72. Operating Sys. & Comp. Arch.II(CS10)			36-41
73. Principles of Database Design			42-47

*'78 refers to courses described in Curriculum '78, *Communications of the Association for Computing Machinery*, Vol. 22, No. 3 (March 1979) 147-166.

**'84 refers to courses described in *Communications of the Association for Computing Machinery*, Vol. 28, No. 8 (August 1985) 815-818.

Appendix B - 3

(Question 3 continued)

Name of Course (or equivalent)	Students Enrolled	Number of Sections	
74. Artificial Intelligence (CS12)			48-53
75. Discrete Structures			54-59
76. Algorithms (CS13)			60-65
77. Software Design & Develop. (CS14)			66-71
78. Principles of Prog. Languages			72-77 / 8
79. Automata, Computability, & Formal Lang. (CS16)			6-11
80. Automata Theory			12-17
81. Numerical Math.: Analysis (CS17)			18-23
82. Numerical Methods			24-29
83. Numerical Math.: Linear Alg.(CS18)			30-35
84. Compiler Design			36-41
85. Networks			42-47
86. Modelling & Simulation			48-53
87. Computer Graphics			54-59
88. Semantics & Verification			60-65
89. Complexity			66-71
90. Computational Linguistics			72-77 / 9
91. Senior Seminar/Independ. Stud. CS			6-11
92. Other Computer Science			12-17

3.C. TOTAL NO. OF COMPUTER SCIENCE SECTIONS — 18-20

4. Instructional Formats

In the table below are listed five courses from the list of question 3. For each course please enter the number of students taught during the Fall of 1985 in each of the formats listed in column (1). In the last line of the table enter the total enrollment in each of these courses in the Fall of 1985. If a course was not offered by your department during this time, enter zero. The numbers in line 7 should be the numbers reported in 3 above.

Number of Students Enrolled: Fall of 1985

(1)	Coll. Alge. (5) (2)	Calculus: (Math., Eng. Phys. Sci.) (15) (3)	Calculus (Bio., Soc., Mgmt. Sci.) (16) (4)	CS1: Comp. Prog. I (56) (5)	Elem. Stat. (45) (6)	
1. Small Class (Less than 40 students)						21-40
2. Large Class (Between 40 and 80 students)						41-60
3. Lecture (Over 80 students) without recitation or quiz sections						61-80
4. Lecture (Over 80 students) with recitation or quiz sections						10 / 6-25
5. Self Paced Instruction						26-45
6. Other Format						46-65
7. Total enrollment in course, Fall '85						66-85

5. Use of Computers

Indicate the number of sections in the courses listed below in which the use of computers (micros/minis/mainframes) is required. (Courses are selected from question 3 where you listed total sections offered.)

	Number of Sections	
A) College Algebra (5)		6-7
B) Calculus (Math., Phys. Sci., Engineering) (15)		8-9
C) Differential Equations (17)		10-11
D) Discrete Mathematics (18)		12-13
E) Linear Algebra and/or Matrix Theory (19)		14-15
F) Numerical Analysis (37)		16-17
G) Elementary Statistics (45)		18-19

(marker 11)

Appendix B - 4

6. Questions on Mathematical and Computer Science Faculty, Fall 1985

A. Report the number of full-time mathematical and computer science faculty members at or above the rank of assistant professor in your department in the table below, by highest degree and subject field in which it was earned:

Highest Degree \ Subject Field	Math.	Stat.	Comp. Sci.	Math. Ed.	Another field	
Doctor's degree						20-29
Master's degree						30-39
Bachelor's degree						40-49

B. Other full-time faculty: Report the number of other full-time faculty in your department, e.g., instructors, not counted in (A), by highest degree and subject field in which it was earned:

Highest Degree \ Subject Field	Math.	Stat.	Comp. Sci.	Math. Ed.	Another field	
Doctor's degree						50-59
Master's degree						60-69 / 12 6-15
Bachelor's degree						

C. How many part-time faculty do your have in your department? ___ 16-17

D. How many of the total sections in each of Mathematics, Statistics, and Computer Science (see question 3) were taught this Fall by the faculty members in A, B, and C of this question?

	No. of Sections Taught in			
	Mathematics	Statistics	Computer Science	
Full-time Assistant Professor or above (A)				18-26
Other Full-time (B)				27-35
Part-time (C)				36-44

Totals of columns should be totals from questions 3A, 3B, 3C, respectively.

E. Departmental Graduate Teaching Assistants (If none, check here ___ and go on to 7.) — 45

  (a) Total number of teaching assistants in Fall, 1985 ___ 46-48

  (b) Number who are graduate students in your department ___ 49-51

  (c) Number who are graduate students in some other mathematical or computer science department. ___ 52-54

F. Use of Departmental Graduate Teaching Assistants (GTA's)

Give the number of your GTA's by their principal teaching responsibility:

  (a) Teaching their own classes ___ 55-56

  (b) Conducting quiz sections or recitation sections ___ 57-58

  (c) Paper grading ___ 59-60

  (d) Providing tutorial or other individual assistance to students ___ 61-62

  (e) Other or no assigned duties ___ 63-64

7. Age, Sex, and Ethnic Group of Full-time Faculty, Fall, 1985

Age	Under 30	30-34	35-39	40-44	45-49	50-54	55-59	60 & Over	
Tenured, Doctorate									55-80 / 13 6-21
Tenured, Non-doctorate									
Non-tenured, Doctorate									22-37
Non-tenured, Non-Doctorate									38-53
Men									54-69 / 14 6-21
Women									
Amer. Indian/Alaskan native									22-37
Asian/Pacific Islander									38-53
Black (not of Hispanic orig.)									54-69
Hispanic									70-85 / 15 6-21
White (not of Hispanic orig.)									

8. A. What is the expected (or typical) teaching load in credit hours for your full-time mathematical science and computer science faculty and GTA's (excluding thesis supervision):

	Mathematical Science Per semester or quarter	Computer Science Per semester or quarter	
(a) Professors			22-25
(b) Associate Professors			26-29
(c) Assistant Professors			30-33
(d) Instructors with Ph.D.			34-37
(e) Instructors without Ph.D.			38-41
(f) Grad. Teaching Ass'ts.			42-45

B. If there are significant departures from these expected teaching loads for certain classes of individuals, please describe:

9. Employment and Mobility of Departmental Faculty

A. Are there any new full-time mathematical science or computer science faculty in your department this year? Yes _____ No _____ — 46

If yes, during the previous year 1984-85 how many were:

	Those with Ph.D.'s	Those without Ph.D.'s	
(1) enrolled in graduate school			47-50
(2) faculty in an institution of higher ed.			51-54
(3) holding postdoctoral study/research appointments			55-58
(4) employed in non-academic positions			59-62
(5) otherwise occupied			63-66

B. Of your full-time mathematical science and computer science faculty last year, are there any who are no longer part of your full-time faculty?
Yes _____ No _____. If yes, how many: (Use the category that best applies.) — 67

	Those with Ph.D.'s	Those without Ph.D.'s	
(1) died, or retired			68-71
(2) were visiting your department and returned to their regular department			72-75
(3) are teaching in an institution of higher education			76-79 / 16 / 6-9
(4) left for a non-academic position			10-13
(5) returned to graduate school			14-17
(6) are otherwise occupied			

11

10. Availability of new faculty members for 1985-86. Please complete the following table.

	In Math.	In Stat.	In Comp. Sci.	
How many full-time openings did you have for faculty for 1985-86?				18-23
How many of these openings were you able to fill with full-time faculty members who met the advertised qualifications?				24-29
How many of these openings did you fill with part-time faculty who otherwise met the advertised qualifications?				30-35
How many of these openings did you fill with faculty who did not meet the advertised qualifications?				36-41
How many of the openings were you unable to fill?				42-47

11. How many bachelor's degrees with majors in a mathematical or computer science were awarded by your department between July 1984 and June 1985? (48-50)

Report the number of bachelor's degrees with major in: Estimate number of those listed on left having at least the equivalent of a minor in:

Major	Comp. Sci.	Math.	Stat.	
Math. (general)		xxxxxx	xxxxxx	51-60
Applied Math.		xxxxxx	xxxxxx	61-70
Math. Ed.		xxxxxx	xxxxxx	71-80 / 17
Comp. Sci.	xxxxxx			6-15
Statistics		xxxxxx	xxxxxx	16-25
Operations Research				26-35
Joint Comp. Sci. & Math.	xxxxxx	xxxxxx		36-45
Joint Math. & Stat.	xxxxxx	xxxxxx	.	46-55
Joint Comp. Sci & Stat.	xxxxxx		xxxxxx	56-65

12. Professional Activities

Below are listed some professional activities which departments may take into account in making recommendations on faculty advancement and/or salary decisions. Please rate each of the items on the 0-5 scale given by encircling a number, with 5 indicating that it is very important for most faculty in your department and is heavily weighted in advancement and/or salary decisions and 0 indicating little or no effect in advancement and/or salary decisions.

	Scale							
no importance	0	1	2	3	4	5 (very important)		

Scale			
0 1 2 3 4 5	A.	Years of service	7
0 1 2 3 4 5	B.	Classroom teaching performance	8
0 1 2 3 4 5	C.	Textbook writing	9
0 1 2 3 4 5	D.	Giving talks at professional meetings	10
0 1 2 3 4 5	E.	Expository and/or popular articles	11
0 1 2 3 4 5	F.	Published research	12
0 1 2 3 4 5	G.	Service to department and/or university (college)	13
0 1 2 3 4 5	H.	Professional activities (participating in work of professional societies, public/gov't services in a professional capacity, etc.)	14
0 1 2 3 4 5	I.	Undergraduate and/or graduate advising	15
0 1 2 3 4 5	J.	Supervision of graduate students	16

13. Problems of the mid 80's

Below are some concerns cited by many departments. Please rate each of the concerns on the 0 to 5 scale given by encircling a number, with 5 indicating a major and continuing problem for your department and 0 indicating no present problem.

Scale			
no problem		major problem	
0 1 2 3 4 5	A.	Losing full-time faculty to industry/gov't	17
0 1 2 3 4 5	B.	Maintaining vitality of faculty	18
0 1 2 3 4 5	C.	Advancing age of tenured faculty	19
0 1 2 3 4 5	D.	Lack of experienced senior faculty	20
0 1 2 3 4 5	E.	Teaching load of full-time faculty	21
0 1 2 3 4 5	F.	The need to use temporary faculty for instruction	22
0 1 2 3 4 5	G.	Salary levels/patterns	23
0 1 2 3 4 5	H.	Promotion/tenure process above department level	24

Scale			
no problem		major problem	
0 1 2 3 4 5	I.	Lack of quality of undergraduate majors	25
0 1 2 3 4 5	J.	Lack of quantity of undergraduate majors	26
0 1 2 3 4 5	K.	Lack of quality of departmental grad. students	27
0 1 2 3 4 5	L.	Lack of quantity of departmental grad. students	28
0 1 2 3 4 5	M.	Class size	29
0 1 2 3 4 5	N.	Remediation	30
0 1 2 3 4 5	O.	Library: holdings, access, etc.	31
0 1 2 3 4 5	P.	Research funding	32
0 1 2 3 4 5	Q.	Departmental support sources (travel funds, staff, secretary, etc.)	33
0 1 2 3 4 5	R.	Computer facilities for faculty use	34
0 1 2 3 4 5	S.	Access to networking facilities for research and communication	35
0 1 2 3 4 5	T.	Upgrading/maintenance of computer facilities	36
0 1 2 3 4 5	U.	Computer facilities for classroom use	37
0 1 2 3 4 5	V.	Office/lab facilities	38
0 1 2 3 4 5	W.	Classroom/lab facilities	39

If you have found some question(s) difficult to interpret or to secure data for, please supply elucidating comments or suggestions which would be helpful to the Committee in future surveys:

If there are any concerns that you would like to discuss in more detail, please include them on a separate sheet.

Information supplied by:

Title and Department: _____

Institution and Campus: _____

Telephone: _____ Date: _____

REMEDIAL MATHEMATICS QUESTIONNAIRE

(For Coding Only)

1. Give the name of the academic unit (or division) administering the remedial (developmental) mathematics program at your institution.

   a) _____ (If not the mathematics dept., please answer b and c.) | 25-26

   b) Give time when unit was established

      ___ Before 1975   ___ 1975-79   ___ 1980-85 | 27

   c) Does the unit report to the same academic administrator as does the mathematics department?   ___ Yes   ___ No | 28

2. Do you have follow-up studies on success rates of students in post-remedial math courses or on eventual graduation:

   ___ Yes (Give name of contact person. _____) | 29
   ___ No

3. For which standard course(s) do remedial mathematics courses prepare students?

   ___ (a) College Algebra   ___ (d) Finite Mathematics | 30-31
   ___ (b) Elementary Functions   ___ (e) Other (Specify name: ____) | 32-33
   ___ (c) Precalculus | 34

4. Staff qualifications and status (for remedial program only).

   (a) Number of full-time faculty | 35-36
   (b) Number of part-time faculty | 37-38
   (c) Number of full-time faculty on tenure track | 39-40
   (d) Number of full-time faculty who are tenured | 41-42

5. Give numbers for full- and part-time faculty (combined) who staff the remedial mathematics program:

Highest Degree \ Field of Degree	Mathematics	Math. Ed.	OTHER	
Doctor's				43-48
Master's				49-54
Bachelor's				55-60
No degree				61-66

Please see reverse side.

---

6. Credit status of remedial courses. Please complete the following table:

Course Title	Is course load credit normally given? YES	NO	Is credit toward graduation given? YES	NO	
Arithmetic					67 / 68
General Math. (Basic Skills, Operations)					69 / 70
Elementary Algebra (High School)					71 / 72
Intermediate Algebra (High School)					73 / 74
Other					75 / 76

| 2

7. Remedial Course Enrollments: (If your unit filled out the main questionnaire, columns (2) and (3), rows (a) to (d), are from Question 3A there.)

	Total No. of Students Enrolled Fall, 1985 (2)	Total No. of Sections (3)	No. Sections Taught by Part-time Faculty (4)	
(a) Arithmetic				7-14
(b) General Math. (Basic Skills, Operations)				15-22
(c) Elementary Alg. (High School)				23-30
(d) Intermed. Alg. (High School)				31-38
(e) Other				39-46

* * *

Information supplied by: _____ Title & Dept.: _____

Institution & Campus: _____   Phone: _____

Date: _____

Please return completed questionnaire by 27 November 1985 to:
Conference Board of the Mathematical Sciences
1529 Eighteenth Street, N.W.
Washington, D.C. 20036

APPENDIX C

TWO-YEAR COLLEGE QUESTIONNAIRE
(SEE PAGE C-6 FOR REMEDIAL QUESTIONNAIRE)

SURVEY OF PROGRAMS IN MATHEMATICS

IN

TWO-YEAR COLLEGES

1985

General Instructions

This questionnaire should be completed by the person who is directly in charge of the mathematics program at your institution.

You are asked to report on all the courses and faculty in your institution which fall under the general heading of the mathematical or computer sciences except for remedial programs taught in a special unit outside the mathematics department. For some colleges this may involve courses and faculty in statistics, applied mathematics and computer science that, although mathematical in nature, are taught outside the mathematics department. If your institution does not have a departmental or divisional structure, consider the group of all mathematics and computer science professors to be the "mathematics department" for the purpose of this questionnaire. Question III below refers to courses taught in the "mathematics department" as explained above. Question IV refers to mathematics and/or computer science courses taught outside the "mathematics department" but not courses taught in a special unit for remediation. Courses in a special unit for remediation taught outside the mathematics department should be reported by the head of that unit in the special questionnaire on remediation (blue page.) Please include data on part-time and evening students and faculty as well as data on occupational and terminal programs. Include non-credit and remedial courses. Do not, however, include data concerning campuses jurisdictionally separate from yours, if such exist.

If the mathematics department offers the remedial program, then the person in charge of the mathematics department should fill in and return the special remediation questionnaire. If another unit offers the remedial program, then the person in charge of that unit should fill out and return the special remediation questionnaire which will be sent to him/her following receipt of the return postcard.

Please return completed questionnaire by 27 November 1985 to:

Conference Board of the Mathematical Sciences
1529 Eighteenth Street, N.W.
Washington, D.C. 20036
(202) 387-5200
* * *

I. A. NAME OF INSTITUTION _____

If this two-year institution is part of a larger organization, identify this relationship: _____

B. Total institutional enrollment Fall 1985 (approximate):

	College Transfer Program		Occupational/Technical	
	Full-time	Part-time	Full-time	Part-time
		Students		Students
Freshman				
Sophomores				
Unclassified or other				
Total				

For Coding Only

1

_ 20

21-40
41-60
61-80
7-26

II. How is the mathematics program administered at your institution?

___ Math. Department   ___ No dept. structure

___ Math. & Comp. Sci. Dept.   ___ Other (specify): _____

___ Math. & Sci. Dept. or Div. | 27 |

III. COURSES IN THE MATHEMATICAL AND COMPUTER SCIENCES OFFERED BY YOUR MATHEMATICS DEPARTMENT in the Fall 1985.

Instructions for preparing table on this and the following page.

A. The courses in column (1) in the following table are listed with typical course titles (which may not necessarily coincide with the titles you use). Additional blank spaces are provided to permit you to write in names of courses which do not fit reasonably under some listed title. Please use your best judgment as to how courses should be listed.

For the purpose of this survey, consider as a single course, instruction in a particular area of mathematics which you offer as a sequence of two or more parts (e.g., calculus.)

B. For each course in column (1) that is offered during Fall 1985, write in column (2) the total number of students who enrolled (in any part of) the course in the Fall term of 1985. Thus, for a 2 semester sequence of calculus, Math 1 and Math 2, the enrollment in column (2) would be the sum of the Fall enrollment in Math 1 and in Math 2. If a course is not being taught in the Fall of 1985, but was offered in the academic year 1984-85 or is expected to be offered during some other term of the 1985-86 academic year, write S for "Sometimes." If a course is not offered during this period, write N for "Not Offered."

C. In column (3) give the total number of sections of the course in Fall 1985.

D. In column (4) give the total number of sections of this course taught by part-time faculty.

E. In column (5) give the total number of sections of this course for which a hand calculator is recommended.

F. In column (6) give the total number of sections of this course in which computer homework assignments are given.

NOTE: EACH BOX IN ANY ROW SHOULD CONTAIN AN ENTRY IF COURSE IS OFFERED IN FALL 1985. EACH BOX IN COLUMN (2) SHOULD CONTAIN A NUMBER, AN N, OR AN S.

Name of Course (or equivalent) (1)	Total No. of Students Enrolled Fall 1985 (2)	Total No. of Sections (3)	No. Sect. Taught by Part-time Faculty (4)	No. Sect. Hand Calc. Recom- mended (5)	No. Sect. Computer Assignments are Given (6)	
1. Arithmetic						28-39
2. General Mathematics (basic skills, operations)						40-51
3. Elementary Algebra (High School)						52-63
4. Intermediate Algebra (High School)						64-75
5. High School Geometry						7-18
6. College Algebra						19-30
7. Trigonometry						31-42

Name of Course (or equivalent) (1)	Total No. of Students Enrolled Fall 1985 (2)	Total No. of Sections (3)	No Sect. Taught by Part-time Faculty (4)	No. Sect. Hand Calc. Recom- mended (5)	No. Sect. Computer Assignments are Given (6)	
8. College Algebra and Trigonometry, Combined						43-54
9. Elem. Functions						55-66 / 67-78
10. Analytic Geometry						7-18
11. Analytic Geometry and Calculus						19-30
12. Calculus (math., phys. sci., & eng.)						31-42
13. Calculus (bio., soc., & mgt. sciences)						43-54
14. Differential Equations						55-66
15. Linear Algebra						67-78
16. Discrete Math.						7-18
17. Finite Math.						19-30
18. Math. for Liberal Arts						31-42
19. Math. of Finance						43-54
20. Business Mathematics						55-66
21. Math. for Elem. School Teachers						67-78
22. Elementary Statistics						7-18
23. Probability (and Stat.)						19-30
24. Technical Mathematics (calculus level)						31-42
25. Technical Math.						43-54
26. Use of Hand Calculators						55-66
27. Data Processing, Elem. or Adv'd						67-78
28. Elem Program'g (BASIC, FORTRAN, PASCAL, COBOL)						7-18
29. Advanced Programming						19-30
30. Assembly Language Programming						31-42
31. Data Structures						43-54
32. Other Comp. Sci. Courses						55-66
33. Other Math. Courses						

4

IV. OUTSIDE ENROLLMENTS - FALL 1985

This question identifies courses in mathematics or computer science taught in divisions or departments of your institution other than that division or department having primary responsibility for mathematics and other than a unit concerned primarily with remedial mathematics. Enter in the relevant boxes an estimate of the total course enrollments for Fall 1985. Please consult schedules to give good estimates of numbers of enrollments. Please enter 0 (zero) in each box for which there are no courses given.

Course	Enrollment in courses given by division specializing in:				
	Natural Sciences	Occupational Programs	Business	Social Sciences	Other
1. Arithmetic						[8] 7-26
2. Business Mathematics						27-46
3. Statistics/ Probability						[9] 47-66
4. Pre-calculus College Math.						7-26
5. Calculus or Diff. Equations						27-46
6. Computer Sci. & Program'g						[10] 47-66
7. Data Processing						7-26
8. Technical Mathematics						27-46
9. Other						47-66

V. QUESTIONS ON MATHEMATICS FACULTY

A. Full-time faculty: indicate in the table below the numbers of your full-time mathematical and computer sciences faculty members teaching courses reported in III above, according to their highest degrees and subject fields in which these were earned:

Highest Degree / Subject Field	In Math.	In Stat.	In Computer Science	In Math. Ed.	In another field	
Ph.D.						[11] 7-16
Ed.D.						17-26
Dr. Arts						27-36
Master's degree, plus 1 year						37-46
Master's degree						47-56
Master's degree (spec. program) e.g., MAT, MST						57-66
Bachelor's degree						67-76

5

B. Part-time faculty: In the table below, indicate the numbers of your part-time faculty by highest degrees and subject fields:

Highest Degree / Subject Field	In Math.	In Stat.	In Computer Science	In Math. Ed.	In another field	
Ph.D.						67-76
Ed.D.						[12] 7-16
Dr. Arts						17-26
Master's degree, plus 1 year						27-36
Master's degree						37-46
Master's degree (spec. program) e.g., MAT, MST						47-56
Bachelor's degree						57-66

C. What is the expected (or typical) teaching load in classroom contact hours for members of your full-time faculty? ___ 67-68

D. How many of your full-time faculty teach overloads? ___ 69-70

E. What is the average overload (in contact hours) for those faculty? ___ 71-72

F. Are faculty normally paid extra for overloads? ___ yes ___ no 73

G. What is the average teaching load in contact hours of part-time faculty? ___ 74-75

H. Of your part-time faculty, how many are: [13]

Employed Full-time in							
High School	In Your Own College	Another Two-year College	Four-year College	Industry or Other	Graduate Students	Not Grad. Students & Not Employed Full-time Anywhere	Total No. of Part-time Faculty
a	b	c	d	e	f	g	t

NOTE: You should have t = a + b + c + d + e + f + g

7-22

VI. USE OF COMPUTERS

A. Does your department have convenient access to a computer or to computer terminal facilities: ___ yes ___ no   23

B. How many of your full-time faculty know a computer language? ___ 24-25

C. How many of your full-time faculty give class assignments involving the use of the computer each year (in courses other than computer sciences)? ___ 26-27

D. How many computer terminals and/or microcomputers are available for student use in your courses? ___ 28-30

6

VII. INSTRUCTIONAL FORMATS

A. At your institution, please indicate the extent to which the following formats are employed. Place a check in the appropriate column.

	Is not Being Used	Is Used by Some Faculty	Is Used By Most Faculty	
1. Standard lecture - recitation system (Class size < 40)				31
2. Large lecture classes (>40) with recitation sections				32
3. Large lecture classes (>40) with no recitation				33
4. Organized program of independent study				34
5. Courses by television (closed circuit or broadcast)				35
6. Courses by film				36
7. Courses by programmed instruction				37
8. (CAI) Courses by computer-assisted instruction				38
9. Modules				39
10. Audio-tutorial				40
11. PSI (Personalized Systems of Instruction)				41
12. Other				42

B. MATH LABS

1. Does your institution operate a math lab or math help (tutorial) center? _____ yes _____ no — 43
   (If you answered no in 1, go on to VIII.)
2. Year math lab was established. Before 1975 ____ , 1975-79 ____ , 1980- ____ — 44
3. Personnel of the math lab include (check all relevant categories):
   ___ Full-time members of the Math. staff   ___ Members of another dept. — 45 _46
   ___ Part-time members of the Math. staff   ___ Paraprofessionals — 47 _48
   ___ Students   ___ Other — 49 _50

7

VIII. COORDINATION OF PROGRAMS

A. Coordination with secondary schools

How often does your math staff consult with secondary schools on development and/or coordination of offerings?

Less Than Once A Year	Yearly	More Than Once Per Year	
			51

B. Coordination with four-year institutions

1. Are your course offerings and/or curriculum subject to state-wide system control or approval? _____ yes _____ no — 52

2. Is there official state-wide coordination of your mathematical offerings with those of four-year state institutions? _____ yes _____ no — 53

3. How often does your institution consult with the mathematics department of four-year colleges on course offerings designed for transfer credit? — 54

Less Than Once A Year	Yearly	More Than Once Per Year

4. Are there other coordination activities involving your mathematics staff and mathematics departments of four-year colleges or universities in your area? _____ yes _____ no — 55

   If yes, please describe these briefly:

IX. FACULTY EMPLOYMENT AND MOBILITY

A. How many of your full-time faculty members were newly appointed on a full-time basis this year? — 56-57

Of this number, during the previous year 1984-85, how many were:

With Doctorate (math.)	With Doctorate (math.ed.)	With Other Doctorate	With No Doctorate		
				1. enrolled in grad. school	58-61
				2. teaching in a 4-year coll. or univ.	62-65
				3. teaching in another 2-year instit.	66-69
				4. teaching in a secondary school	70-73
				5. employed by you part-time	74-77 / 14
				6. employed in non-academic position	7-10
				7. otherwise occupied or unknown	11-14

B. Of the full-time faculty in 1984-85 who are no longer part of your full-time faculty, how many:

	With Doctorate (math.)	With Doctorate (math.ed.)	With Other Doctorate	With No Doctorate	
1. died, or retired					15-18
2. are teaching in a 4-year coll. or univ.					19-22
3. are teaching in a 2-year institution					23-26
4. left for a non-academic position					27-30
5. returned to graduate school					31-34
6. left for secondary school teaching					35-38
7. are otherwise occupied or unknown					39-42

X. AGE, SEX, AND ETHNIC GROUP OF FULL-TIME FACULTY

Record the number of full-time faculty members in each category:

AGE	<30	30-34	35-39	40-44	45-49	50-54	55-59	≥60	
Bachelor's									43-58
Master's									59-74 / 15
Doctor's									7-22
Men									23-38
Women									39-54
Amer. Indian/Alaskan native									55-70 / 16
Asian/Pacific Islander									7-22
Black (not of Hispanic orig.)									23-38
Hispanic									39-54
White (not of Hispanic orig.)									55-70

XI. PROFESSIONAL ACTIVITIES

Estimate the number of full-time members of your department who, in a given year,

1. ____ attend at least one professional meeting — 71-72
2. ____ take additional mathematics or computer science courses — 73-74
3. ____ attend minicourses or short courses — 75-76
4. ____ give talks at professional meetings — 77-78
5. ____ regularly read articles in professional journals — 79-80 / 17
6. ____ write expository and/or popular articles — 7-8
7. ____ publish research articles — 9-10
8. ____ write textbooks — 11-12

9

XII. PROBLEMS OF THE MID-'80's

Below are some concerns cited by many departments. Please rate each of the concerns on the 0 to 5 scale given by encircling a number, with 5 indicating a major and continuing problem and 0 indicating no present problem.

	no problem		Scale			major problem	
A. Losing full-time faculty to industry/gov't	0	1	2	3	4	5	13
B. Maintaining vitality of faculty	0	1	2	3	4	5	14
C. Advancing age of tenured faculty	0	1	2	3	4	5	15
D. Lack of experienced senior faculty	0	1	2	3	4	5	16
E. Staffing computer science courses	0	1	2	3	4	5	17
F. The need to use temporary faculty for instruction	0	1	2	3	4	5	18
G. Salary levels/patterns	0	1	2	3	4	5	19
H. Class size	0	1	2	3	4	5	20
I. Remediation	0	1	2	3	4	5	21
J. Library: holdings, access, etc.	0	1	2	3	4	5	22
K. Departmental support sources (travel funds, staff, secretary, etc.)	0	1	2	3	4	5	23
L. Computer facilities for faculty use	0	1	2	3	4	5	24
M. Upgrading/maintenance of computer facilities	0	1	2	3	4	5	25
N. Computer facilities for classroom use	0	1	2	3	4	5	26
O. Office/lab facilities	0	1	2	3	4	5	27
P. Classroom/lab facilities	0	1	2	3	4	5	28
Q. Coordinating and/or developing math. for vocational/technical programs	0	1	2	3	4	5	29
R. Coordinating math. courses with 4-year colleges and universities	0	1	2	3	4	5	30
S. Coordinating math. courses with high schools	0	1	2	3	4	5	31

* * *

Information supplied by: _____

Title: _____

Telephone: (_____) _____ _____ — 32-33
Area Code Number Extension

1. How long have you been in charge of the mathematics program? _____ years — 34
2. Is the chairmanship rotating? _____ yes _____ no — 35
   If yes, what is the frequency of rotation? _____
3. If you have found any of the above survey questions difficult to interpret or to secure data for, please supply elucidating comments or suggestions which would be helpful to the Committee in future surveys:

REMEDIAL MATHEMATICS QUESTIONNAIRE

For Coding Only

1. Give the name of the academic unit (or division) administering the remedial (developmental) mathematics program at your institution.

   a) _____ (If not the mathematics dept., please answer b and c.)

   b) Give time when unit was established
   ___ Before 1975 ___ 1975-79 ___ 1980-85 — 27

   — 25-26

   c) Does the unit report to the same academic administrator as does the mathematics department? ___ Yes ___ No — 28

2. Do you have follow-up studies on success rates of students in post-remedial math courses or on eventual graduation: — 29

   ___ Yes (Give name of contact person. _____)
   ___ No

3. For which standard course(s) do remedial mathematics courses prepare students?

   ___ (a) College Algebra   ___ (d) Finite Mathematics   — 30-31
   ___ (b) Elementary Functions ___ (e) Other (Specify name: _____)   — 32-33
   ___ (c) Precalculus   — 34

4. Staff qualifications and status (for remedial program only).

   ___ (a) Number of full-time faculty   — 35-36
   ___ (b) Number of part-time faculty   — 37-38
   ___ (c) Number of full-time faculty on tenure track   — 39-40
   ___ (d) Number of full-time faculty who are tenured   — 41-42

5. Give numbers for full- and part-time faculty (combined) who staff the remedial mathematics program:

Highest Degree \ Field of Degree	Mathematics	Math. Ed.	OTHER	
Doctor's				43-48
Master's				49-54
Bachelor's				55-60
No degree				61-66

6. Credit status of remedial courses. Please complete the following table:

Course Title	Is course load credit normally given? YES	NO	Is credit toward graduation given? YES	NO	
Arithmetic					— 67 / — 68
General Math. (Basic Skills, Operations)					— 69 / — 70
Elementary Algebra (High School)					— 71 / — 72
Intermediate Algebra (High School)					— 73 / — 74
Other					— 75 / — 76

7. Remedial Course Enrollments: (If your unit filled out the main questionnaire, columns (2) and (3), rows (a) to (d), are from Question 3A there.)   ☐ 2

	Total No. of Students Enrolled Fall, 1985 (2)	Total No. of Sections (3)	No. Sections Taught by Part-time Faculty (4)	
(a) Arithmetic				7-14
(b) General Math. (Basic Skills, Operations)				15-22
(c) Elementary Alg. (High School)				23-30
(d) Intermed. Alg. (High School)				31-38
(e) Other				39-46

* * *

Information supplied by: _____ Title & Dept.: _____

Institution & Campus: _____ Phone: _____

Date: _____

Please return completed questionnaire by 27 November 1985 to:

Conference Board of the Mathematical Sciences
1529 Eighteenth Street, N.W.
Washington, D.C. 20036

Please see reverse side.

APPENDIX D

SPECIAL QUESTIONNAIRE ON COMPUTER SCIENCE

SPECIAL COMPUTER SCIENCE QUESTIONNAIRE

This part of the questionnaire is designed for those departments which offer under-graduate programs (not necessarily degree programs) in computer science. It is limited to courses in computer science and the faculty which teach them. It is intended to give more detailed information about computer science itself than that recorded in the general survey. Summary information on your faculty has been included in the main questionnaire. All questions refer to Fall 1985 data. For computer science departments, per se, a few of the questions may be duplicates.

For Coding Only

1. A. Which of the following subject areas best describes the Computer Science bachelor's degree(s), if any, offered by your department (check as many as apply):

___ No bachelor's degree in computer science	___ Business	_ _ 25-26
___ Science (Liberal Arts and Science)	___ Education	_ _ 27-28
___ Engineering	___ Other	_ _ 29-30

B. Which, if any, departments or units (other than your own) on your campus teach undergraduate computer science courses (check as many as apply):

___ Mathematics	___ Library	_ _ 31-32
___ Engineering	___ Humanities	_ _ 33-34
___ Business	___ Education	_ _ 35-36
___ Other Natural Science	___ Computer Center	_ _ 37-38
___ Social Science		_ 39

2. A. Percentage of students enrolled in departmental Computer Science courses with programming projects using:

	Micros	Minis/Mainframes	Total	
In lower level courses in 3C of main questionnaire			100%	40-45
In middle or upper level courses in 3C of main questionnaire			100%	46-51

B. Percentage of work stations used in departmental Computer Sciences courses controlled by:

	Department	Non-Department	Total	
Micros			100%	52-57
Minis/Mainframes			100%	58-63

C. Consider the number of students taking departmental computer science courses and using the computer in Fall 1985. Check the average number of student enrollments per work station.

0-5 ___ ; 6-10 ___ ; 11-15; ___ ; 16-20 ___ ; 21 or more ___ .   _ 64

2

3. A. Of the non-computer science courses listed in 3A of the main questionnaire, encircle (by code numbers in 3A) those required for computer science majors.

15 16 17 18 19 20 21 22 23 24 25 26 27 28 29 30 31 32 33 34

35 36 37 38 39 40 41 42 43 44 45 46 47 48 49 50 51 52 53 54   7-46

B. What is the total number of mathematics (and statistics) semester or quarter courses (at the calculus level and above) normally taken by computer science majors? _____   47-48

Please see reverse side.

4. A. <u>Full-time</u> Computer Science faculty. Report the number of full-time computer
   science faculty in your department in the table below, by the highest degree
   and subject field in which it was earned (if the number is zero, check here _____):
   (The numbers should total to your full-time computer science faculty.)   _ 49

Highest degree Field	CS	Stat	Math	Educ	Engin	Other
Doctor's degree						
Master's degree						
Bachelor's degree						

50-61

62-73
[3
6-17

B. <u>Part-time</u> Computer Science faculty, other than your teaching assistants.
   Report the number of faculty teaching Computer Science part-time in your department
   in the table below, by highest degree and subject field in which it was earned:
   (If the number is zero, check here _____ .)   _ 18

Highest degree Field	CS	Stat	Math	Educ	Engin	Other
Doctor's degree						
Master's degree						
Bachelor's degree						

19-24

25-30

31-36

C. Of the part-time computer science faculty reported in 4B above, how many were

   ____ (a) Employed full-time by your university or college   37-38
   ____ (b) Employed full-time by some other university or college   39-40
   ____ (c) Employed full-time by a high school   41-42
   ____ (d) Employed full-time but not in education   43-44
   ____ (e) Not employed full-time anywhere   45-46

D. How many of the full-time and part-time departmental faculty reported in 4A and 4B
   teach:

	Full-time	Part-time
(a) basically only lower level courses?		
(b) only specialty courses?		

47-52

53-58

E. Of the faculty reported in 4A and 4B above, how many have joint appointments in
   Computer Science and:

   ___ Mathematics   ___ Other Natural Sciences   59-60
   ___ Engineering   ___ Social Sciences   61-62
   ___ Business   ___ Humanities   63-64
   ___ Library Science   ___ Research Institutes   65-66
               * * *   ___ Other   _ 67

Information supplied by:_____ Title & Dept.:_____

Institution & Campus:_____ Phone:_____

Date:_____

<u>Please return completed questionnaire by 27 November 1985 to:</u>

   Conference Board of the Mathematical Sciences
   1529 Eighteenth Street, N.W.
   Washington, D.C. 20036

APPENDIX E

COURSE BY COURSE ENROLLMENTS IN UNIVERSITIES AND FOUR-YEAR COLLEGES (In Thousands)
Sums may not total, because of rounding.
(L means some but less than 500)

Name of Course (or equivalent)	Universities	Public Colleges	Private Colleges	Total
A. MATHEMATICS				
Remedial				
1. Arithmetic	3	8	4	15
2. General Math. (basic skills, operations)	2	18	11	31
3. Elem. Algebra (High School)	15	52	8	75
4. Intermed. Alg. (High School)	36	77	17	130
Total Remedial	56	155	40	251
Pre-calculus				
5. College Algebra	53	73	25	150
6. Trigonometry	12	22	3	37
7. College Alg. & Trig. combined	31	35	12	78
8. Elem. Functions, Pre-calc. Math.	26	30	18	74
9. Math. for Liberal Arts	12	30	17	59
10. Finite Mathematics	35	30	23	88
11. Business Mathematics	12	22	3	37
12. Math. for Elem. School Teachers	12	31	10	54
13. Analytic Geometry	L	2	1	3
14. Other Pre-calculus	7	5	1	13
Total (Non-remedial) Pre-calc.	200	280	113	593
Calculus Level				
15. Calc. (Math., Phys. Sci. & Eng.)	162	163	77	402
16. Calc. (Bio., Soc. & Mgmt. Scis.)	73	49	8	129
17. Differential Equations	22	18	6	45
18. Discrete Mathematics	5	8	2	14
19. Linear Alg. and/or Matrix Theory	19	20	8	47
Total Calculus	281	258	101	637
Advanced Level				
20. Modern Algebra	5	6	2	13
21. Theory of Numbers	1	1	1	3
22. Combinatorics	2	3	L	4
23. Graph Theory	1	L	L	1
24. Coding Theory	-	L	-	L
25. Foundations of Mathematics	1	2	1	3
26. Set Theory	L	1	-	1
27. Discrete Structures	1	3	3	7

Name of Course (or equivalent)	Universities	Public Colleges	Private Colleges	Total
A. MATHEMATICS				
Advanced Level (Continued)				
28. History of Mathematics	L	1	-	2
29. Geometry	2	4	1	7
30. Math. for Secondary School Teachers (Methods, etc.)	1	3	1	5
31. Mathematical Logic	1	L	1	2
32. Advanced Calculus	5	6	3	14
33. Advanced Math. for Eng. & Physics	4	5	1	10
34. Vector Analysis. Linear Algebra	4	8	2	14
35. Advanced Diff. Equations	1	2	L	4
36. Partial Diff. Equations	1	3	-	5
37. Numerical Analysis	5	7	2	13
38. Applied Mathematics, Math. Modelling	1	1	1	4
39. Operations Research	3	2	1	6
40. Complex Variables	2	2	1	5
41. Real Analysis	2	2	2	5
42. Topology	1	L	L	2
43. Senior Seminar/Independ. Stud. Mathematics	L	1	1	2
44. Other Mathematics	3	3	1	7
Total Advanced Level	47	66	25	138
B. STATISTICS				
45. Elem. Stat. (no Calc. prereq.)	40	41	34	115
46. Probability (& Stat.) (No Calc. prerequisite)	12	13	5	29
47. Mathematical Statistics (Calc.)	10	9	6	24
48. Probability (Calculus)	7	5	3	15
49. Stochastic Processes	L	-	-	L
50. Applied Stat. Analysis	7	3	1	11
51. Design & Analysis of Experiments	1	L	-	1
52. Regression (and Correlation)	1	L	-	1
53. Senior Seminar/Indep. Stud. Stat.	L	-	-	L
54. Other Statistics	11	1	L	12
Total - All Statistics	89	72	49	208
C. COMPUTER SCIENCE				
Lower Level				
55. Computers & Society	10	36	23	69
56. CS1 '78 or CS1 '84 (Computer Programming I)	36	50	43	129
57. CS2 '78 (Computer Prog. II)	6	13	8	28

Name of Course (or equivalent)	Universities	Public Colleges	Private Colleges	Total
C. COMPUTER SCIENCE				
Lower Level (Continued)				
58. CS2, '84	4	7	4	15
59. Database Mgmt. Systems	1	4	2	7
60. Discrete Mathematics	3	8	2	12
61. Other Lower Level Service	34	37	19	90
Total Lower Level	94	155	101	350
Middle Level				
62. Intro. to Comp. Systems (CS3)	4	11	3	18
63. Assembly Lang. Programming	6	13	5	24
64. Intro. to Comp. Organization	5	6	3	14
65. Intro. to File Processing (CS5)	3	4	2	10
Total Middle Level	18	34	13	66
Upper Level				
66. Operating Sys. & Computer Arch.	2	1	1	4
67. Operating Systems	4	5	2	11
68. Computer Architecture	2	2	2	6
69. Data Structures (CS7)	7	10	7	24
70. Survey of Prog. Languages	3	5	1	9
71. Computers & Society (CS9)	L	L	L	1
72. Operating Systems & Comp. Architecture II (CS10)	1	1	L	2
73. Principles of Database Design	3	2	2	7
74. Artificial Intelligence (CS12)	3	1	1	5
75. Discrete Structures	2	2	1	4
76. Algorithms (CS13)	2	3	–	5
77. Software Design & Develop.(CS14)	3	3	2	8
78. Principles of Prog. Languages	2	3	1	6
79. Automata, Computability, & Formal Languages (CS16)	2	2	L	4
80. Automata Theory	1	1	–	2
81. Numerical Math.: Analysis (CS17)	1	2	1	4
82. Numerical Methods	1	1	L	2
83. Numerical Math: Linear Alg. (CS18)	1	1	1	2
84. Compiler Design	2	2	–	4
85. Networks	1	1	1	3
86. Modelling & Simulation	L	1	L	1
87. Computer Graphics	2	2	1	6
88. Semantics & Verification	L	L	–	L
89. Complexity	L	L	–	L
90. Computational Linguistics	–	–	–	–
91. Senior Seminar/Independ. Stud. CS	2	1	1	4
92. Other Computer Science	7	9	3	18
Total Upper Level	54	61	28	142

APPENDIX F

LIST OF RESPONDENTS TO SURVEY

A: Public Universities

Arizona State University	Decis. & Info. Sci., Mathematics Computer Science
Bowling Green State University	Applied Stat. & Oper. Res., Computer Science, Mathematics & Statistics
Indiana State University	Mathematics & Computer Science
Indiana University - Bloomington	Computer Science
Iowa State University	Computer Science, Statistics, Mathematics
Michigan State University	Mathematics, Computer Science, Statistics & Probability
New Mexico State University	Exper. Statistics, Mathematical Sciences, Computer Science
North Dakota State University	Mathematical Sciences
North Texas State University	Computer Science
Northern Illinois University	Mathematical Sciences
Ohio State University	Computer & Info. Sciences, Statistics
Ohio University-Athens	Computer Science, Mathematics
Oklahoma State University-Stillwater	Mathematics, Statistics
Penn State University-University Park	Statistics, Mathematics, Computer Science
SUNY at Buffalo	Mathematics, Statistics
South Dakota State University	Mathematics, Computer Science
Texas A & M University	Computer Science, Mathematics, Statistics
University of Akron	Mathematical Sciences
University of Arizona	Statistics
University of Cincinnati	Mathematics, Computer Science
University of Delaware	Mathematical Sciences
University of Florida	Mathematics, Statistics, Computer and Informational Systems
University of Georgia	Mathematics, Mathematics Education, Computer Science, Statistics
University of Idaho	Mathematics, Computer Science
University of Kentucky	Mathematics, Statistics
Univ. of Maryland-College Park	Mathematics, Computer Science
Univ. of Michigan-Ann Arbor	Mathematics, Statistics
Univ. of Minnesota-Minneapolis	Computer Science, Mathematics, Statistics

Univ. of Missouri-Columbia	Computer Science, Statistics Mathematics
Univ. of Nebraska-Lincoln	Mathematics & Statistics, Computer Science
University of Nevada	Mathematics
University of New Hamshire	Mathematics, Computer Science
University of New Mexico	Computer Science
Univ. of North Carolina-Chapel Hill	Mathematics, Biostatistics
University of Oregon	Mathematics
Univ. of Puerto Rico-Rio Piedras	Mathematics
University of South Dakota	Mathematical Sciences
University of Texas-Austin	Mathematics
University of Vermont	Mathematics & Statistics
University of Wisconsin-Madison	Statistics, Mathematics
Washington State University	Pure & Applied Mathematics, Computer Science
Wichita State University	Mathematics and Statistics, Computer Science

B. Private Universities

Boston University	Mathematics, Computer Science
Carnegie-Mellon University	Statistics, Mathematics
Drake University	Mathematics & Computer Science
Duke University	Computer Science, Mathematics
George Washington University	Statistics & Comp. Inf. Systems, Mathematics, Operations Research
Georgetown University	Mathematics, Computer Science
Johns Hopkins University	Mathematics, Electrical Eng. & Computer Science, Mathematical Sciences
New York University	Computer Science
Northeastern University-Boston	Mathematics
Northwestern University-Evanston	Ind. Eng. & Management Sciences, Mathematics
Seton Hall University	Mathematics
Syracuse University	Comp. & Info. Sciences, Mathematics
Texas Christian University	Mathematics, Computer Science
University of Miami	Mathematics & Computer Science
University of Santa Clara	Mathematics
Univ. of Southern California	Mathematics
University of Tulsa	Mathematics & Computer Science
Vanderbilt University	Mathematics
Wake Forest University	Mathematics & Computer Science
Washington University	Computer Science

C. Public Four-Year Colleges

Alabama A & M University	Mathematics
Bluefield State College	Nat. Sciences
Boise State University	Mathematics
CUNY-Brooklyn College	Computer & Info. Science, Mathematics
CUNY-City College	Mathematics
CUNY-Hunter College	Computer Science
CUNY-Queens College	Computer Science, Mathematics
Cal Maritime Academy	Mathematics
Cal Polytech State University	Statistics, Computer Science Mathematics
Cal State College-Stanislaus	Mathematics
Cal State Polytech University	Mathematics
Cal State University-Chico	Mathematics
Cal State University-Long Beach	Mathematics & Computer Science
Cal State University-Northridge	Mathematics
Cameron University	Mathematics
Central Michigan University	Computer Science, Mathematics
Central State University	Computer Science
Chadron State College	Mathematics & Science
Christopher Newport College	Computer Science, Mathematics
Cleveland State University	Mathematics, Comp. & Info Sci.
College of William & Mary	Computer Science, Mathematics
Delta State University	Mathematics
East Central Oklahoma St. Univ.	Computer Science, Mathematics
Eastern Illinois University	Mathematics
Eastern Oregon State College	Mathematics & Computer Science
Florida International University	Mathematical Sciences
Francis Marion College	Math. & Computer Sciences
George Mason University	Mathematical Sciences, Systems Engineering, Computer Science
Georgia Institute of Technology	Mathematics
Harris-Stowe State College	Mathematics & Science
Indiana-Purdue Univ.-Ft. Wayne	Mathematical Sciences
Indiana-Purdue Univ.-Indianapolis	Mathematical Sciences, Computer & Info. Sciences
Jacksonville State University	Mathematics
Lamar University	Mathematics
Longwood College	Mathematical & Computer Sciences
Mississippi Univ. for Women	Science & Mathematics
Missouri Southern State College	Computer Science, Mathematics
New Mexico Inst. Mining/Tech.	Mathematics, Computer Science
North Carolina A & T State Univ.	Mathematics & Computer Science
Northern Montana College	Mathematics
Northern State College	Mathematics
Northwestern Oklahoma State Univ.	Computer Science
Pan American University	Mathematics & Computer Science
Penn State Univ.-Capitol Campus	Mathematical Sciences
Plymouth State College	Mathematics, Computer Science
Ramapo College of New Jersey	Mathematics
Rutgers University-Newark	Mathematics

SUNY College at Buffalo	Mathematics
SUNY College at Cortland	Mathematics
SUNY College at Fredonia	Mathematics & Computer Science
SUNY College at Geneseo	Mathematics
SUNY College at Oswego	Mathematics
SUNY College at Tech	Math.-Art & Sciences
	Computer & Info. Sciences
SUNY Maritime College	Mathematics
SUNY at Stony Brook	Computer Science, Applied Mathematics & Statistics, Mathematics
Saginaw Valley State College	Science & Engineering
San Diego State University	Mathematical Sciences
San Jose State University	Mathematics & Computer Science
Sangamon State University	Mathematics
South Dakota School Mines & Tech.	Mathematical Sciences
Southeast Missouri State University	Mathematics, Computer Science
Southern Ill. Univ.-Edwardsville	Mathematics-Stat.-Comp. Sci.
St. Cloud State University	Mathematics & Computer Science
Stephen F. Austin State University	Mathematics & Statistics
Texas Southern University	Mathematics
Troy State University-Ft. Rucker	Mathematics
Univ. of California-San Diego	Mathematics
Univ. of California-Santa Cruz	Mathematics
University of D.C.	Mathematics
University of Illinois at Chicago	Mathematics
University of Minnesota-Duluth	Mathematical Sciences
University of Missouri-St. Louis	Mathematical Sciences
University of North Florida	Mathematical Sciences
University of Puerto Rico-Humacao	Mathematics
University of South Florida	Mathematics, Computer Science & Engineering
University of Texas-Arlington	Mathematics
University of Wisconsin-Eau Claire	Computer Science, Mathematics
University of Wisconsin-Milwaukee	Mathematics
University of Wisconsin-Platteville	Mathematics, Computer Science
Valley City State College	Mathematics
Washburn University of Topeka	Math. & Info. Sciences
Western Carolina University	Mathematics & Computer Science
Western Michigan University	Computer Science, Mathematics
Western Washington University	Computer Science

D. **Private Four-Year Colleges**

Antillian College	Physical Science
Bellevue College	Mathematics
Bentley College	Mathematics
Berea College	Mathematics
Blue Mountain College	Mathematics
Brown University	Applied Mathematics, Mathematics
Bryant College	Mathematics

California Lutheran College	Mathematics-Physics-Comp. Sci.
Cardinal Stritch College	Mathematics & Computer Science
Carleton College	Mathematics
Central College	Mathematics & Computer Science
Chatham College	Mathematics, Info. Science
Coe College	Mathematics & Computer Science
College of Idaho	Mathematics
Colorado Technical College	Mathematics
DePauw University	Mathematics & Computer Science
Edgewood College	Mathematics & Computer Science
Fairleigh Dickinson Univ.-Teaneck	Mathematics & Computer Science
Florida Institute of Technology	Mathematics & Computer Science
Florida Memorial College	Science & Mathematics
Gordon College	Mathematics & Computer Science
Hamilton College	Mathematics & Computer Science
Heidelberg College	Mathematics
Hood College	Mathematics
Illinois College	Computer Science
Iona College	Mathematics, Comp. Info. Science
Jarvis Christian College	Mathematics
Keuka College	Mathematics
La Salle University	Mathematical Sciences
Lakeland College	Mathematics & Computer Science
Manchester College	Mathematical Sciences
Marion College	Mathematics, Computer Science
Midland Lutheran College	Mathematics
Millikin University	Mathematics & Computer Science
New Hampshire College	Mathematics
Oberlin College	Mathematics
Oklahoma Christian College	Mathematics
Oral Roberts University	Mathematics & Science
Our Lady of Holy Cross College	Natural Sciences
Pace University	Mathematics
Rider College	Mathematics & Physics
Rochester Institute of Technology	Mathematics
Samford University	Math-Engr-Computer Science
Siena College	Mathematics
Southern College	Mathematics
St. John's University	Computer Science
St. Joseph's University	Mathematics & Computer Science
St. Leo College	Science & Mathematics
St. Mary's College	Computer Science
Sterling College	Applied Mathematics
Talladega College	Mathematics
Trinity Christian	Mathematics
University of Bridgeport	Mathematics
University of Dayton	Computer Science
University of San Francisco	Mathematics
University of Steubenville	Mathematics-Computer Science
Virginia Wesleyan College	Mathematics & Computer Science
West Coast Univ.-Orange City Ctr.	Computer Science, Arts & Sciences
Westbrook College	Mathematics
York College of Pennsylvania	Mathematics-Physical Sci.

E. Two-Year Colleges

Alexander City State Jr. College	Mathematics
Alpena Community College	
Amarillo College	Mathematics & Engineering
American River College	Mathematics & Engineering
Austin Community College	Mathematics & Phys. Science
Bakersfield College	Mathematics
Bergen Community College	Natural Sci. & Mathematics
Brevard Community College	Mathematics
Broward Community College	Math.-Cen. Campus
Burlington County College	Sci.-Math.-Technology
Butte Community College	Mathematics
CUNY-New York City Tech. College	Mathematics
Central Texas College	Mathematics
Cerritos College	Mathematics
Chabot College	Science & Mathematics
Cloud County Community College	
Coastline Community College	
College of the Albemarle	
Community College of Beaver County	
Contra Costa College	Mathematics
Cuyahoga Community College	
Cypress College	Science Engr. & Mathematics
Davenport College of Business	
DeKalb Community College	Math. Eng. & Computer Science
El Camino College	Science & Mathematics
El Paso Community College	Mathematics and Science Div.
Erie Community College	Mathematics
Everett Community College	Mathematics
Fresno City College	Mathematics & Science
Gainesville Jr. College	Mathematics
Galveston College	Mathematics & Natural Science
Harrisburg Area Community College	Math. Eng. & Technology
Haywood Technical College	
Henry Ford Community College	Mathematics
Imperial Valley College	Math. Eng. & Science
Inter American Univ.-Aquadilla,PR	Mathematics
Jacksonville College	Mathematics
Johnston Technical College	
Joliet Junior College	
Kellogg Community College	Science & Mathematics
Kirtland Community College	Mathematics
Lackawanna Jr. College	
Lake City Community College	Mathematics
Lake Region Community College	Mathematics
Lake-Sumter Community College	Math-Sci-All'd Health
Lakeland Community College	Mathematics
Lane Community College	Mathematics
Lansing Community College	Mathematics
Laredo Jr. College	Mathematics
Lincoln College	Mathematics

Long Beach City College	Mathematics & Engineering
Los Angeles City College	Mathematics
Los Angeles Pierce College	Mathematics
Los Angeles Southwest College	Mathematics
Los Angeles Trade Tech. College	Mathematics & Science
Los Angeles Valley College	Mathematics
Louisiana State Univ.-Alexandria	Mathematics
Macomb Community College	
Merced College	Mathematics & Science
Montgomery College-Takoma Pk.	Mathematics
Moorpark College	Mathematics
Morristown College	Mathematics
Mt. San Antonio College	Mathematics & Astronomy
Muskegon Community College	Mathematics & Science
North Harris City College	Mathematics
Northeast Alabama St. Jr. College	Mathematics
Ocean County College	Mathematics
Odessa College	Mathematics
Ohio State Univ.- Agri. Tech. Inst.	
Olympic College	Mathematics
Orange Coast College	Mathematics & Phys. Science
Oxnard College	Mathematics & Science
Pima Community College	
Portland Community College	Mathematics
Prince George's City Community College	Mathematics & Engineering
Rancho Santiago College	Mathematics
Rapphannock Community College	
Richard Bland College	Mathematics & Statistics
Ricks College	Mathematics
Rock Valley College	Mathematics & Humanities
Rockingham Community College	Mathematics & Science
San Antonio College	Mathematics
San Diego City College	
San Jacinto College	Mathematics & Engineering
Santa Monica College	Mathematics
Santa Rosa Jr. College	Mathematics
Scottsdale Community College	Mathematics & Science
Southwestern Michigan College	Mathematical Sciences
Spartanburg Tech. College	
Surry Community College	Mathematics & Science
Tallahassee Comm. College	Mathematics & Science
Tarrant County Jr. College	Mathematics
Temple Jr. College	Mathematics
Texas State Tech. Inst.-Amarillo	Mathematics
Tidewater Community College	Mathematics
Tri-Cities State Tech. Inst.	
Tulsa Jr. College	Mathematics
Vernon Regional Jr. College	Mathematics & Science
Villa Julie College	
Vista College	Mathematics & Science
Wayne Community College	Mathematics
Wayne County Community College	

Wenatchee Valley College   Mathematics
West Los Angeles College   Mathematics & Science
Western Piedmont Community College
Wilkes Community College
William R. Harper College   Tech-Mathematics-Phys. Science
York Technical College   Mathematics